НЕКОТОРЫЕ ВОПРОСЫ ФИЗИКИ ПЛАСТИЧНОСТИ КРИСТАЛЛОВ

NEKOTORYE VOPROSY FIZIKI PLASTICHNOSTI KRISTALLOV

PLASTICITY OF CRYSTALS

PLASTICITY OF CRYSTALS

edited by

M. V. Klassen-Neklyudova

Authorized translation from the Russian

CONSULTANTS BUREAU
NEW YORK
1962

Library of Congress Catalog Card Number 62-12853

The Russian text was published by the USSR Academy of Sciences
Press for the All-Union Institute of Scientific and Technical
Information of the Academy of Sciences of the USSR in Moscow
in 1960 as Volume 3 of the Physical-Mathematical Sciences
Series of *Science Summaries (Itogi Nauki)*

CONTENTS

PUBLISHER'S NOTE

The following Russian Journals cited in this book are available in cover-to-cover translation

Russian title	English title	Publisher
Doklady Akademii Nauk SSSR	Soviet Physics — Doklady	American Institute of Physics
Fizika tverdogo tela	Soviet Physics — Solid State	American Institute of Physics
Izvestiya Akademii Nauk SSSR; Seriya fizicheskaya	Bulletin of the Academy of Sciences of the USSR: Physical Series	Columbia Technical Translations
Kristallografiya	Soviet Physics — Crystal-lography	American Institute of Physics
Metallurgiya i topliva	Russian Metallurgy and Fuels	Eagle Technical Publications
Uspekhi fizicheskikh Nauk	Soviet Physics — Uspekhi	American Institute of Physics
Zavodskaya laboratoriya	Industrial Laboratory	Instrument Society of America
Zhurnal éksperi-mental'noi i teoreticheskoi fiziki	Soviet Physics — JETP	American Institute of Physics
Zhurnal tekhnicheskoi fiziki	Soviet Physics — Tech-nical Physics	American Institute of Physics

M. V. KLASSEN-NEKLYUDOVA

PHYSICAL BASIS OF PLASTICITY AND STRENGTH OF CRYSTALS (INTRODUCTION)

The successful development of any branch of science depends on the level of the theoretical knowledge. Up to the present time there has been no strict physical theory of the plasticity and strength of crystals. The strength and resistance to plastic deformation of crystals and polycrystals at different temperatures and rates of deformation are still determined experimentally for each particular substance. This is rather complicated, since the characteristics depend not only on the chemical composition of the substance but also on the surface state of the samples, their previous history, the surrounding media, etc.

During the 1920's and 1930's there arose the necessity of creating a physical theory of the mechanical properties of crystals and, first of all, monocrystals.

This problem, presented to the physicists by the engineers, was first noted in the classical work by V. L. Kirpichev. In the Soviet Union the basic work necessary for the development of the physics of the strength and plasticity of solids had begun, immediately after the October Revolution, with the work of A. F. Ioffe, I. V. Obreimova, N. N. Davidenkov, V. D. Kuznetsov, P. A. Rebinder, S. T. Konobeevskii, and N. Ya. Selyakov.

I. A. Ioffe and M. V. Kirpicheva applied the x-ray method of investigation to the study of the nature of structural transformations induced in crystals by plastic deformation for the first time in the history of science (1922). In 1925, I. V. Obreimov developed a method of growing very large monocrystals of metals and a method of investigating stresses in deformed (transparent) crystals by using polarized light. These large crystals of metals and other compounds made it possible to study the anisotropy of their mechanical properties in terms of their atomic structure. The x-ray and polarization-optical methods made it possible to clarify the nature of phenomena determining the change of the shape of crystals as the result of stress, and the nature of destruction processes. In view of the investigation by N. N. Davidenkov concerning the testing of materials, and the theoretical work of Ya. I. Frenkel', we can say that as early as the 1920's there existed in the Soviet Union the necessary foundation for the successful development of the physics of the strength and plasticity of solids.

1

However, in the USSR as well as abroad, this line of investigation was not only slowed down but stopped by the extraordinary progress and prospects of nuclear physics, which for some time obscured the fundamental problems of the physics of solids. During and after the war the general problems of the physics of solids, and particularly the physics of strength and plasticity, were continued in the Soviet Union by only a few enthusiasts, to whom it was always clear that regardless of the problems posed by science the strength of materials would always limit the solution of any problem and that the development of the physical theory of the strength of materials could not be removed from the program of study. In recent years there has been a considerable increase of interest in this problem in the United States and England. The number of reports on the mechanical properties of crystals has multiplied.

In the Soviet Union interest in the physics of the strength of crystals has also increased, and the problem of the "physical basis of the strength and plasticity of crystals" has been included among the most important problems by the Academy of Sciences. In the USSR the development of research along this line has been retarded by the lack of specialties in this field and by the lack of literature available in the Russian language. In recent years Read's Dislocations in Crystals has been translated (under the editorship of I. A. Oding) and also Cottrell's Dislocations and Plastic Flow in Crystals in 1958 (under the editorship of A. G. Rakhshtadt). However, it should be emphasized that these books were published abroad as early as 1953 and since then the number of investigations in this field increased at such a rate that it became necessary to print a collection (Russian translation) of the most interesting theoretical and experimental work published after the two books mentioned. This collection, called Dislocations in Crystals, was published in 1957 under the editorship of A. N. Orlov. In the preface S. V. Vonsovskii and A. N. Orlov indicate the place occupied by each of the articles in the extensive literature, and also mention the most interesting investigations made later in the USSR and abroad which were not included in the collection. The main achievements in the theory of dislocation in recent years have been published in the review by V. L. Indenbom, which contains summaries of investigations of individual dislocations, their groupings, and their displacement, made with optical and electronic microscopy, selective etching, decoration, the study of moiré patterns, etc. [1].*

The present collection is the work of the personnel of the Laboratory of the Mechanical Properties of Crystals of the Institute of Crystallography of the Academy of Sciences USSR during 1957-1958. It is an attempt at a sys-

---

* V. L. Indenbom and A. N. Orlov have written a review, "Physical theory of plasticity and strength," which is to be published in Uspekhi Fizicheskikh Nauk No. 2 or No. 3 (1962).

tematic description of the present problems of the physics of plasticity and strength, and to a certain extent is meant to fill the gap in the Russian literature on these problems.*

This collection contains articles devoted to the plastic properties of monocrystals and one article devoted to the description of the latest views and experimental data on the structure and properties of grain boundaries. The first four articles concern some problems of the plastic properties of one crystal grain (crystallite). The last article is a review of the basic studies of the mosaic (block) structure of monocrystals and grains of polycrystals, and also the structures and properties of interlayers between grains and aggregates.†

Almost all the articles of this collection contain results of original investigations by the authors of the articles and the basic conclusions are based on original data as well as data in the literature.

The first article, by V. R. Regel', is an investigation of the effect of temperature and rate of deformation on resistance to plastic deformation (flow limit) and work hardening. These questions are important in the development of the physical theory of heat resistance and creep of metals. The processes involved in the flow of polycrystalline metals at high temperatures and low deformation rates are very complex and depend, first of all, on the flow within grains and along grain boundaries. An enormous number of investigations have been devoted to the behavior of industrial materials at high temperatures. The present collection contains only an investigation of the flow of monocrystals, i.e., single grains. The role of flow along the grain boundaries is not dealt with in this article. The validity of treating the plasticity of crystals as a process not dependent on temperature is discussed. The historical development of the theory is given, including the attempt to provide a theory of the temperature and time dependence of the plastic properties of crystals based on the concept of dislocation.

V. R. Regel' discusses the experimental results concerning crystals and amorphous bodies and juxtaposes two current theories of the plasticity and strength of solids: the dislocation theory of crystals, and the activation theory of amorphous bodies. Regel' considers the attempt to develop the dislocation theory of amorphous bodies justified. The first attempt in this direction was the work of Frank, Keller, and O'Connor [2].

In the next two articles A. A. Urusovskaya reviews the experimental data on the character of structural transformations of crystals during plastic flow.

---

* In 1960 an extensive monograph (in English) by Van Bueren was published in Holland: "Imperfections in crystals" [10].
† For the English translation of this book the author has added notes and remarks which to a certain extent reflect the results of investigations made in the past three to four years.

The purpose of the first of these reviews is to clarify, on the basis of x-ray data, the difference between the real process of slip in crystals and the classical scheme of translational slip proposed by Mugge some time ago. The cause of the appearance of asterism in Laue diagrams is discussed and cases of plastic deformation not inducing the elongation of Laue spots are described This information is necessary for further development of the theory of the mechanism of plastic flow in crystals. In the second review Urusovskaya examines the experimental data on plastic flow which differs from ordinary slip either when the distribution of stresses in samples is not homogeneous (mono- and polycrystals) or when deformation by ordinary slip is rendered difficult by the orientation of the crystals or grains. Different complicated processes of plastic deformation resulting from rotated regions are described.*

The article by V. L. Indenbom is a review of the theoretical investigation based on the assumption that a collective displacement of atoms in the process of plastic flow of crystals can be reduced to the motion of a special type of defect— dislocation in crystals. Assuming that the experimental data accumulated so far indicate that dislocations in crystals really exist, and possess properties predicted by the theory, the author considers the possibility of dislocational descriptions of different characteristic phenomena of plastic deformation. The experiments made by Obreimov, Nye, Stepanov, Regel', and the author in collaboration with Tomilovskii and others are analyzed on the basis of the assumption that the translational slip can be reduced to the motion of dislocation in the slip plane, accompanied by the compression of the crystal on one side of the slip plane and elongation along the other. Indenbom calculates the latent energy and stress due to the presence of dislocations in the slip plane and also the effective surface energy of the slip plane. He shows that the reorientation of the lattice regions in stressed crystals can be represented by the motion and regrouping of dislocations. The results of the oretical investigations of the formation of rotated blocks, deformation bands, faults bands, and twins are compared with the experimental data. Of particular interest is the description of the Ball and Hirsh general theorem of the relationship between the active elements of slip and the character of disorientation of blocks in the process of plastic deformation, and also the treatment of the dislocation theory of the relationship between the elements of slip and the elements of classic twinning presented by Bullough. The last section of this review gives examples of the interpretation of some general concepts of the microscopic theory of elasticity and plasticity in dislocational terminology. This shows the possibility of presenting the dislocation theory in terms of the microscopic theory of the mechanism of displacement of

---

* This problem is considered in greater detail in a monograph by M. V. Klassen-Neklyudova [11].

atoms. In particular the author derives an equation for the plane which is a generalization of the basic equations of the theory of plastic bending [37].

The last article of this collection, by V. F. Miuskov, reviews the latest concepts of the structure of intergranular boundaries.* Miuskov describes the foundation of the dislocation theory of the structure of the grain boundaries, mosaic blocks, and subgrains. He demonstrates the possibility of the quantitative determination of the degree of disorientation of neighboring blocks by the formulas of the dislocation theory on the basis of the density of etch pits formed along the block boundary. The last part of this article is devoted to experimental investigations which confirm the conclusions of the dislocation theory of boundaries on the basis of the investigation of a number of basic properties of intercrystalline interlayers. The basic properties of boundaries are investigated in terms of the disorientation angle between touching grains. The motion of boundaries in a field of mechanical stresses, their confluence, and their separation into their initial components are described. It is clear that the present model of dislocational structure of the grain boundaries is not completely justified. In fact we now have a model of the structure of intercrystalline interlayers. This model correctly reflects the basic property of the interlayers, and the theory makes it possible to arrive at sufficiently precise calculations of the surface energy of the block boundaries, and even grain boundaries,† and also predicts the previously known properties of interlayers (e.g., the anisotropy of diffusion along the grain boundary, the existence of which is confirmed by experiments).

Furthermore, on the basis of the relationship between the properties of intercrystalline boundaries and the disorientation angle between touching grains, it was possible to obtain important new data on the segregation of impurities along the boundaries, on the phenomenon of the fusion of boundaries, and on the microscopic displacement of grains along the boundaries. It was observed that the microscopic motion of grains along the boundaries is jumplike and the heat of activation of this process depends on the disorientation angle in a particular way. These data are very important for clarifying the process of creep along grain boundaries.

In fact it is possible to describe different types of defects in crystals by use of dislocation models. As to the dislocation theory of the plasticity and strength of crystals, at the present time the theory is incomplete. Theoretical investigations in this area are only outlines of the theory of single phenomena. In the preface to the American edition of Read's book (1953) he wrote: "It is a disappointment that I can offer only fragments of a theory of

---

* A review of this problem by Amelinckx and Dekeyser appeared in 1959 [12].

† In spite of the fact that the formulas of the dislocation theory of boundaries are correct only for very small disorientation angles.

mechanical properties, strength, plastic deformation, and work hardening—
these are the most important applications of dislocation theory from a practical viewpoint. However, it appears to me more profitable to recognize that
at present there is no systematic general theory that is convincing or widely
accepted." In spite of considerable progress in the development of a theoretical basis for the dislocation hypothesis and considerable experimental confirmation, this assertion by Read is unfortunately correct even now.

Let us indicate, for example, the most important problem concerning
the sources of dislocation. For many years the problem had no rational solution. After 1950, when Frank and Read found a geometrically possible scheme
of multiplication of dislocations—known as the Frank-Read source—it seemed
that the main difficulties of the theory were overcome. By a method of decoration and selective etching of crystals of silicon [4], cadmium [5], zinc[6],
and germanium [7] the Frank-Read sources were actually found. However,
Gilman and Johnston [8] negated the possibility of explaining their experiments on the appearance of dislocations in lithium fluoride crystals on the
basis of Frank-Read sources. Kuhlmann-Wilsdorf [9] also notes a number of
facts contradicting the ordinary assumption that the sources of dislocation
created during the process of plastic deformation are segments of a dislocational net acting as a Frank-Read source. Many authors concluded that the
crystal surface, grain boundaries, and accumulation of point defects of the
lattice play particular roles in the formation of the nuclei of plastic deformation. In particular, the accumulation of vacancies can close up and form
prismatic dislocations, which in turn are capable of acting as sources of dislocations by a mechanism similar to that of the Frank-Read source.

The history of the development of our concepts of slip nuclei shows once
again that a physical theory of plasticity and strength requires strict experimental checking of different possible schemes of the collective displacements
of atoms and critical use of existing theoretical and experimental data.

Moscow, 1958

LITERATURE CITED

1. V. L. Indenbom, "Dislocations in crystals," Kristallografiya 3, No. 1, 13
    (1958).
2. F. C. Frank, A. Keller, and O'Connor, "A deformation process in poly-
    ethylene interpreted in terms of crystal plasticity," Phil. Mag. 3, No. 25,
    64 (1958).
3. W. T. Read, Dislocations in Crystals [Russian translation] (1957).
4. W. C. Dash, "Copper precipitation in silicon," J. Appl. Phys. 27, No. 10,
    1193 (1956).

5. N. A. Tyapunina and A. A. Predvoditelev, "Spiral etching patterns in cadmium polycrystals," Doklady Vysshei Shkoly, Fiz-Mat Nauki No. 6, 184 (1958).

6. S. Servi, "Etching patterns in high-purity zinc," Phil. Mag. 3, No. 25, 63 (1958).

7. W. W. Tyler and W. C. Dash, "Dislocation arrays in germanium," J. Appl. Phys. 28, No. 11, 1221 (1957).

8. J. J. Gilman and W. G. Johnston, "Creation of dislocations in LiF crystals at low stresses," J. Appl. Phys. 29, No. 1 (1958).

9. D. Kuhlmann-Wilsdorf, "On the origin of dislocations," Phil. Mag. 3, No. 26, 125 (1958).

10. H. G. van Bueren, Imperfections in Crystals (North-Holland Publishing Company, 1960).

11. M. V. Klassen-Neklyudova, Mechanical Twinning in Crystals [in Russian] (Izd. AN SSSR, 1960).

12. S. Amelinckx and W. Dekeyser, "The structure and properties of grain boundaries," Solid State Physics 8 (1959).

V. R. REGEL'

TIME AND TEMPERATURE DEPENDENCE
OF PLASTICITY CHARACTERISTICS
IN MONOCRYSTALS

During recent years the effect of time and temperature on the strength and plasticity of solids has become a very important practical problem due to the need for materials capable of withstanding high temperatures and stresses over long periods of time. A great number of theoretical and experimental investigations of this problem have been published in recent years.

Since the problem as a whole is very complex, it is desirable to separate the purely technical problems from the theory of the process of plastic deformation and rupture.

From the whole problem of the effect of time and temperature on the strength and plasticity of solids we have limited this article to the consideration of monocrystals.

To develop a general physical theory of the strength and plasticity of solids it is essential to have experimental data on monocrystals. One of the most important criteria of the validity of the general physical theory of the plasticity of crystals is whether the theory can correctly interpret the existing experimental data on the temperature and time dependence of the strength of crystals. Therefore comparison of the experimental results with the present theoretical considerations is important for the solution of the whole problem.

For monocrystals the basic characteristics of plasticity and strength are usually considered to be the parameters of compression and elongation curves (obtained at a constant deformation rate) and creep curves (obtained under constant stress). Other characteristics derived from more complex stress states are taken into consideration for polycrystals (e.g., from experiments on bending or torsion, determination of hardness, etc.). In the present article we consider only the experimental data concerning the tension and compression of monocrystals under different time and temperature conditions, and we shall discuss these data on the basis of the existing theory of the plasticity of crystals. We shall not deal with the phenomena of the inelastic state (imperfect elasticity) or relaxation.

We have included in the bibliography additional literature on the temperature and time dependence of the characteristics of plasticity and strength not only of monocrystals but also of some polycrystalline materials. This additional bibliography is only a partial list of the studies which have been made.

SOME CONCEPTS OF THE EFFECT OF TIME AND TEMPERATURE ON THE STRENGTH AND PLASTICITY OF AMORPHOUS AND POLYCRYSTALLINE SOLIDS

The temperature and time dependence of the mechanical properties of elastic viscous bodies can be described by the Maxwell equation, which correlates the stress σ and the relative deformation ε:

$$\frac{d\varepsilon}{dt} = \frac{1}{G}\frac{d\sigma}{dt} + \frac{\sigma}{\eta} = \frac{1}{G}\left(\frac{d\sigma}{dt} + \frac{\sigma}{\tau}\right), \tag{1}$$

where G is the shear modulus, η is the viscosity coefficient, and τ is the relaxation time.

This equation makes it possible to explain the experimentally known relationship between the mechanical properties of amorphous substances and the rate of their deformation. The behavior of amorphous bodies during deformation depends on the relationship between the values of the deformation rate, $v = d\varepsilon/dt$, and the rate of plastic deformation, σ/η. When $v \gg \sigma/\eta$ the body behaves as an elastic brittle body, and when $v \ll \sigma/\eta$ it behaves as completely viscous. This description of the time dependence of the plasticity of amorphous bodies is widely accepted (Kobeko [1]).

The temperature dependence of the mechanical properties of elastic viscous bodies which obey Maxwell's equation is, in fact, determined only by the relationship between the viscosity coefficient η, or the relaxation time τ, and the temperature T, since the dependence of G on T is relatively small, and, as a rule, can be neglected.

The mechanism of plastic deformation of amorphous bodies as well as the mechanism of the viscous flow of liquids is in this case related to the process of regrouping of separate atoms or molecules, which can occur under the effect of heat flow. The rate of viscous flow is determined by the probability of such regroupings. The relaxation time is inversely proportional to this probability:

$$\tau = \tau_0 e^{\frac{U}{kT}}, \tag{2}$$

where U is the activation energy of the regrouping process, k is the Boltzmann constant, and τ_0 is a constant.

It is easy to check experimentally whether the relationship between ε and σ in a given body follows Eq. (1) and whether the coefficients of this equation characterizing the mechanical properties of the body, G and τ, are independent of σ and t.

To determine G and τ experimentally it is advantageous to apply testing methods under such conditions that one of the derivatives, $d\varepsilon/dt$ or $d\sigma/dt$, in Eq. (1) will become 0 or at least remain constant. For this reason one uses stress relaxation tests under the condition that σ = const., and compression or elongation tests under the condition that $v = d\varepsilon/dt$ = const. It is easy to establish a correlation between the constants G and τ of Eq. (1) and the parameters of the relaxation curves, creep curves, and deformation curves if v = const.

The experimental data concerning the kinetics of the deformation of real bodies, including simple amorphous bodies, show that the relationship between ε and σ is much more complex than that predicted by Eqs. (1) and (2).

Many authors attempt to correlate the theory with the experiments on the basis of Maxwell's theory of the consecutive combination of elastic and viscous elementary models, but introduce an additional concept concerning the mechanism of regrouping, leading to plastic deformation.

The assumption that the rate of plastic deformation is proportional to the probability of individual regrouping of atoms and molecules simplifies the phenomenon too much. Kobeko [1] indicates that viscous flow of liquids is in fact correlated with the collective displacement of separate layers or large groups of atoms with respect to each other. The conception of the collective character of the regrouping of atoms in the process of plastic deformation of amorphous bodies can also be found in the work of other authors.

The assumption that the potential barrier during regrouping is overcome only as the result of thermal fluctuations is another simplifification. In the process of regrouping the acting stresses accomplish a certain amount of work, which decreases the necessary activation energy. The decrease of the activation energy can be considered, in a first approximation, to be proportional to σ, since the work of external forces during the regrouping period is equal to the force acting on the particles multiplied by the distance between two equilibrium positions. One can therefore assume that in expression (2) U must have the form

$$U = U_0 - a\sigma, \qquad (3)$$

where a is a constant with the dimension of the volume. It follows then that the relaxation time τ which enters into Maxwell's equation depends exponentially on the stress:

$$\tau = \tau_0 e^{\frac{U_0 - a\sigma}{kT}}. \qquad (4)$$

By introducing these concepts Aleksandrov [2] and his collaborators [3, 4] succeeded in explaining the observed relationships during plastic deformation of glassy polymers. According to this concept the plastic deformation of poly-

mers can occur also at low temperatures, i.e., below the softening temperature. The potential barriers are overcome not only because of the fluctuation
of the thermal energy but also because of the work of external forces.

Gurevich [5, 6] proposed applying Maxwell's equation by taking into account the relationship between τ and σ to describe the mechanical properties
of a great number of materials, including polycrystalline materials. By taking
into account the structure and the related hardening, Gurevich proposed to
derive Maxwell's equation by considering U_0 in expressions (3) and (4) to be a
function of the relative deformation ε. This function can have different forms
for different materials, depending on the hardening mechanism. Thus some
investigators use Maxwell's equation, including the relationship between τ and
σ, to describe the kinetics of plastic deformation not only of simple amorphous but also of high molecular and polycrystalline materials.

It must be emphasized that such descriptions of the plasticity of solids do
not require introduction of the concept of critical stresses of the type of the
flow limit or elasticity limit, which determine the conditions for the beginning of plastic flow. According to Eqs. (1) and (4) plastic flow can occur under
any stress. Since the relationship between τ and σ is exponential, the deformation curves may have sharp changes in direction, indicating regions of stress
within which a considerable flow rate is developed, i.e., the flow limit (apparent). However, this indication, useful for practical purposes, has no physical meaning. In fact the real constants of the material are τ_0, U_0 and a.

In recent years Zhurkov and his collaborators have obtained important experimental data on the temperature and time dependence of the strength and
plasticity of amorphous and polycrystalline substances. They have shown [7,
8, 9, 10] that for a wide range of materials the dependence of the strength on
temperature and on the prolonged effect of stress can be well described by an
expression of type (4), where τ is the longevity of the sample. Recently
Zhurkov and Sanfirova [11] showed that the same type of expression also described very well the dependence of the rate of steady creep on σ and T. It
turned out that the activation energy U_0 and the constant a in the expressions
of the rate of creep and longevity have the same value, and that U_0 is numer-
ically equal to the sublimation energy and not to the self-diffusion energy, as
was assumed previously by other investigators, e.g., Dorn [12]. The fact that
in Eq. (4) U_0 and a have identical values in the process of destruction and plastic deformation indicates that there is a close relationship between the two
processes.

The results of the experiment described by Zhurkov and Sanfirova could
be interpreted by Maxwell's Eq. (1), and by expression (4). However, these
authors do not find it necessary to indicate this. At the same time they emphasize that to characterize the processes of destruction and plastic deformation one must not introduce the concept of critical stress of the type of limit

resistance and limit flow since they have no physical meaning. The kinetic character of these processes does not make it possible to characterize them without considering the time factor.

Many authors consider Maxwell's concepts a basis for the explanation of the plastic properties of solids. In those cases when the relationship between τ and σ described above is insufficient to explain the experimental facts, other hypotheses are introduced. Sometimes several different mechanisms of regrouping are assumed to exist in real bodies and, correspondingly, several different relaxation periods, with different values for τ_0, U_0, and a. Some authors give a more complex relationship between the relation time and σ than that given in Eq. (4). In a certain number of investigations the linear dependence of U on σ is replaced by a quadratic relationship. Shestopalov [13] calculates the relaxation of stress in metals on the basis of Maxwell's equation, using for the relationship between τ and σ the following:

$$\tau = \tau_0 e^{\frac{U_0 - U_1 \, \text{th} \, \gamma\sigma}{kT}}$$

(5)

In the general case, instead of Eq. (3) one can write:

$$U = U(\sigma).$$

(6)

The expression of τ becomes more complicated than that given in (4) if we consider not only those regroupings to which stress contributes, leading to the potential barrier (3), but also those for which the acting stresses increase the energy barrier:

$$U = U_0 + a\sigma.$$

(7)

The consideration of regrouping with barriers of the type in (3) and (7) leads to the expression:*

$$\tau = \tau_0 e^{\frac{U_0}{kT}} \, \text{sh} \, \frac{a\sigma}{kT} .$$

(8)

There are also numerous investigations in which an attempt is made to describe the kinetics of deformation by other equations of state (for example, see Zener and Hollomon [15]) and by other schemes. More complex formal schemes represent combinations of parallel and consecutive inclusions of elastic and viscous elements. The corresponding equations are linear combinations of terms containing ε and σ and their first derivatives. Schemes with a parallel introduction of elements are used to explain the imperfect elasticity in metals (Zener [16]), high elasticity in polymers (Lazurkin [4]), and other phenomena. In the present article we shall not consider these problems.

* The same expression can be derived if one considers the process of flow from the standpoint of the theory of the rate of chemical reactions (see for example, the work of Kauzmann [14]).

EXPERIMENTAL INVESTIGATIONS OF THE EFFECT OF TEMPERATURE AND DEFORMATION RATE ON THE FLOW LIMIT AND HARDENING OF MONOCRYSTALS UP TO 1935

The original theory of the plasticity of crystals as well as amorphous bodies was based on the relationship between the relaxational regrouping of atoms and the process of plastic flow.

As we have said before, in such a mechanism of plastic deformation the concept of critical stresses, which determine the beginning of plastic deformation, has no physical meaning.*   At the same time the experimental data on the elongation of metal monocrystals, summarized in the well-known monograph by Schmid and Boas [18], show that for monocrystals it is meaningful to introduce the concept of critical stress, which determined the beginning of the flow of the crystal as the result of slip.

In describing the shape of the elongation curve of cadmium monocrystals, Schmid and Boas indicate that the region of intense hardening (a sharp rise of the stress curve corresponding to an insignificant elongation) is suddenly followed by the region of large elongation corresponding to an insignificant increase in stress. This transition is so sharp that apparently one can ascribe a definite physical meaning to the stress at which such considerable slip begins; this stress is called the flow limit of the crystal. Consequently, the determination of the flow limit in metal monocrystals is based on the process of deformation itself and is not determined, as in the case of polycrystals, by some arbitrary condition.

This reasoning by Schmid and Boas does not appear to us very convincing, since if we take into account the exponential relationship between τ and σ the change in the direction of the deformation curves can be quite sharp even in the case of "amorphous" plasticity. However, to confirm the meaningfulness of the concept of critical stresses for monocrystals,

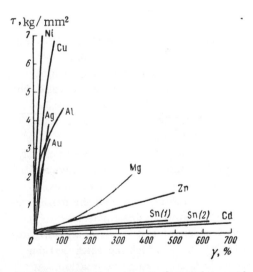

Fig. 1. Hardening curves of metal crystals (according to Schmid and Boas [18]). τ) Shearing stress; γ) displacement in the direction of slip.

---

* Frenkel' came to the same conclusion [17], along with many other authors.

Schmid and Boas give numerous experimental data concerning the effect of orientation, temperature, and deformation rate on the flow limit.

The experimental investigation of the effect of the crystallographic orientation of the sample on the flow limit led Schmid and Boas to conclude that intense slipping begins when the slip system reaches some definite critical shear stress (the law of shear stresses). At the same time the normal stress becomes insignificant. The effect of normal stresses begins to be felt only at high hydrostatic pressures. The investigation of the effect of orientation on the shape of the elongation curve led to the conclusion that all the numerous elongation curves corresponding to different orientations can be well reproduced by a single curve—"the hardening curve"—drawn in crystallographic coordinates: the specific displacement* and shear stress in a slip system.

Some of the curves from the Schmid and Boas monograph are shown in Figs. 1, 2, and 3; the shapes of the hardening curves of some metal monocrystals and the effect of the temperature and deformation rate on the hardening curve and on the critical shear stress are given. Analysis of these and other results from hexagonal and face-centered cubic metals led Schmid and Boas to the following conclusion on the effect of temperature and time on the slipping process: the critical shear stress characterizing the beginning of noticeable slip depends very little on the temperature and deformation rate; at absolute zero the resistance to shear is of the same order of magnitude as at high temperatures.

Contrary to the low temperature dependence of the flow limit, there is a very strong temperature dependence of hardening γ, at least within a certain temperature range. In the range of the lowest temperatures the flow limit and the slope of the hardening curve are first independent of temperature, but with increasing temperature there is a region in which hardening decreases greatly, and finally a temperature region where the hardening curve has a small slope and, again, is very little dependent on temperature.

The effect of the deformation rate on the hardening curve is completely analogous. This effect is vanishingly small at the lowest temperatures and near the melting point, but is quite significant at average temperatures.

On the basis of these facts, Schmid and Boas assumed that the main process is athermic, i.e., not dependent on temperature or deformation rate. It can occur at the lowest temperatures and is determined by the characteristic function of the substance—by the hardening curve, which indicates that the resistance to shear increases considerably with increasing deformation. At higher temperatures this basic athermic process is superimposed on the heat-dependent process of annealing or relaxation of the crystal. Consequently, in the general case of deformation of the crystal by slip, two processes, different in nature, are acting simultaneously: one is induced by the crystal itself

---

* A relative displacement in the direction of slip.

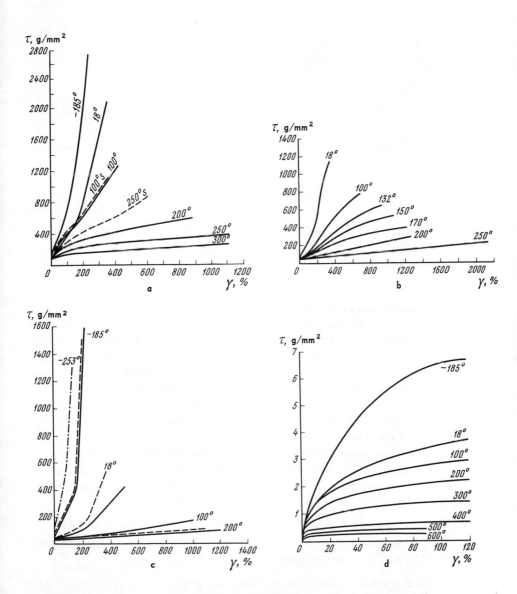

Fig. 2. Effect of temperature on the hardening curves of metal crystals (according to Schmid and Boas [18]). a) Magnesium; b) zinc; c) silicon; d) aluminum; τ) shearing stress; γ) displacement in the direction of slip. The dotted lines represent tests made at a deformation rate 100 times greater.

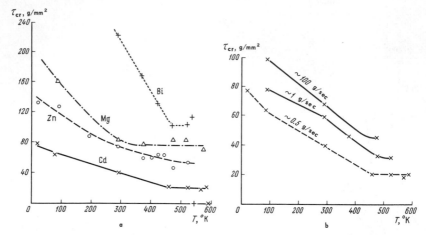

Fig. 3. Effect of temperature T on the critical shear stress of metal monocrystals a, and the same relationship for Cd crystals b, at different rates of stress (according to Schmid and Boas [18]).

and is determined by its structure, and the other is determined by the thermal vibrations of the lattice sites. With increasing temperature the relaxation continually decreases the hardening resulting from slipping. Finally, at temperatures close to melting point, relaxation leads to deformation without hardening; this deformation occurs at a constant sheat stress within a given slip system. This description of the phenomenon of slip also explains the time dependence of these phenomena: an increase of the acting time of the stress (i.e., the decrease of the deformation rate) plays the same role in annealing as an increase of temperature.

The monograph by Schmid and Boas summarizes the results of investigations by many authors over a period of several years, and the conclusions reached were accepted at that time. With this in mind we shall not refer to earlier works of many other authors published before 1935.

RECENT EXPERIMENTAL INVESTIGATIONS OF THE EFFECT OF T AND v ON THE FLOW LIMIT AND HARDENING OF CRYSTALS WITH DIFFERENT STRUCTURES

Publication of the Schmid and Boas monograph was followed by that of additional experimental data which widened our concept of the kinetics of the plastic deformation of crystals.

Stepanov and his co-workers showed [19, 20] that for the same crystal the temperature dependence of the flow limit in the case of slip (σ_T) can be very different along different slip planes. Figure 4 gives data [19, 20] on the effect

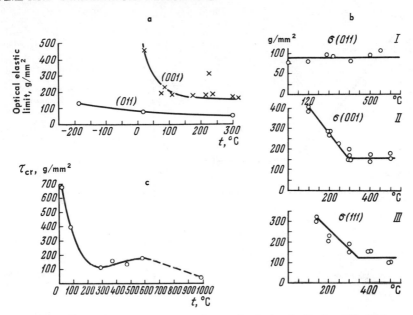

Fig. 4. Effect of temperature on the optical elastic limit of NaCl in different slip planes a (according to Stepanov et al. [19, 20]), b on the flow limit (according to Theile [21]), c.

of T on σ_T for the (110), [110]; (100), [110]; and (111), [110] systems in NaCl crystals. It can be seen that in the first system σ_T is practically independent of T, i.e., the process appears to be athermic; for the other systems this dependence is much more apparent. For comparison we have shown on the same figure the data given by Theile [21], which we took from [18]. Contrary to Stepanov, Theile determines the flow limits for the (110), [110] systems not by the optical method but by the elongation curve.

In a number of investigations concerning the effect of T on body-centered cubic metal crystals it was found that the flow limit of these crystals depends much more on temperature and the rate of deformation than in face-centered cubic or hexagonal close-packed crystals. McAdam and Mebs [22], Vogel and Brick [23], Jokobory [24], and Paxton and Bear [25] investigated α-iron monocrystals. Figure 5 represents the variation of the flow limit of α-iron as a function of temperature, taken from publications [22, 23, and 73]. The curve shows that when the temperature varies from -100 to $+50°C$, σ_T varies approximately by a factor of five. Maddin and Chen [26] obtained a similar relationship between σ_T and T for Mo monocrystals. Some other body-centered cubic metal crystals gave analogous results (partly based on experiments with polycrystals) — Bechtold [27] investigated tungsten, Barrett α-

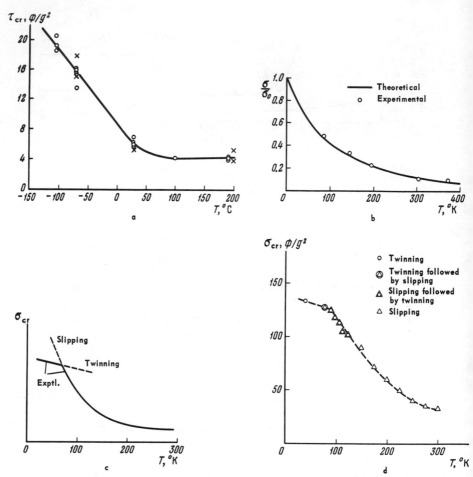

Fig. 5. Effect of temperature on the flow limit of α-Fe crystals. a) Circles represent the results of tensile strength tests while crosses are the results of compression tests (according to Vogel and Brick [23]); b) comparison of the experimental data from McAdam and Mebs [22] with the theoretical data from Cottrell and Bilby [73]; c) schematic curve of the variation of σ_{cr} as a function of T for α-Fe corresponding to the transition of the deformation me- chanims from twinning to slipping at low temperatures (according to Erikson and Low [39]); d) effect of T on σ_{cr} for α-Fe (according to Erikson and Low [39]).

brass [28], Jamison and Sherrill β-brass [29], etc. The effect of the crystal structure on the character of the temperature dependence of the flow limit was also indicated by Davidenkov and Chuchman in their review of the modern theory of low temperature of embrittlement [30]. However, most of the work cited here concerns investigations of polycrystalline metals and not monocrystals.

The considerable dependence of σ_T on T and v was also found in the case of nonmetallic body-centered cubic crystals. Regel', Zemtsov, and Tomilovskii [31-34] investigated the variation of the flow limit of TlBr— TlI monocrystals as a function of time and temperature. These crystals are transparent and have a structure analogous to that of CsI monocrystals; their mechanical properties are close to those of metals. The variation of σ_T as a function of T for these crystals is represented in Fig. 6. The figure shows that the variation of σ_T as a function of T was measured at ten different deformation rates. The two limit rates differ by a factor of about 10^5.

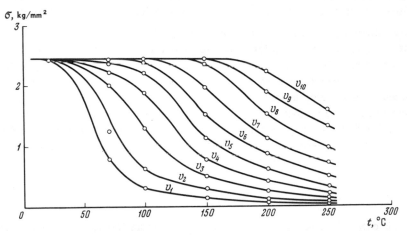

Fig. 6. Effect of temperature and deformation rate on the yield point of TlBr + TlI monocrystals (according to Regel' [34]).

Thus the data on body-centered cubic metals and some other crystals show that the conclusion of Schmid and Boas that σ_T is very little dependent on T, which they derived from investigations of face-centered cubic and hexagonal close-packed metals, is not universal and cannot be applied to crystals of any crystallographic structure or composition. This means that the theory of plasticity of crystals cannot be based on the assumption that the beginning of plastic deformation in any crystal is athermic (independent of temperature and deformation rate). The theory must explain the dependence of σ_T on T and v, at least for some crystals. Experiments show that in a number of cases the degree of dependence of σ_T on T and v is determined by the degree of purity of the crystal. In crystals having a large amount of impurities σ_T is much more dependent on T and v than in pure crystals (e.g., see the previously cited work of Gilman [58]).

Among the investigations of the effect of temperature on the characteristics of plasticity of crystals, those made at temperatures close to absolute zero are very interesting. Blewitt et al. [35, 36], Basinski and Sleesmyk [37], Noggle and Koehler [38], Erikson and Low [39], and others who worked first on low temperature showed that slip in crystals can occur even at very low temperatures and that the flow limit remains finite but has a small value when $T \rightarrow 0°K$. This fact is often used as the basis of the assumption of the difference in the mechanism of deformation of monocrystals and amorphous substances. It has been assumed that amorphous plasticity results from atomic regrouping induced by thermal motion, and that when the temperature approaches absolute zero it must disappear. This assumption could be correct if the variation of the amorphous plasticity as a function of temperature were determined by a formula of the type: $\tau = \tau_0 e^{U/kT}$. In fact this formula cannot be applied to the description of the plasticity of amorphous bodies at low temperatures. Experiments show that amorphous substances undergo plastic deformation even at very low temperatures. Lazurkin [4] has observed a forced elastic deformation* of high molecular substances in the glassy state even at the temperatures of liquid hydrogen and liquid helium. To explain this fact, as we have said, it was assumed that the relaxation time for the regrouping determining plastic deformation depends on the stress, and that this dependence is given by the relationship:

$$\tau = \tau_0 e^{\frac{U_0 - a\sigma}{kT}} \ .$$

In this case τ can be small even at low temperatures provided $U_0 \approx a\sigma$. From this point of view one can speak of a certain analogy between the properties of amorphous high molecular substances and monocrystals since both are subject to plastic deformation at the lowest possible temperatures under the effect of not very large stresses.

A number of investigations concerning the mechanical properties of crystals at low temperatures is devoted to determining the conditions of the transition from plastic deformation to rupture and from slipping to twinning. The transition from one mechanism of deformation to the other may lead to a change in the character of the dependence of σ_T on T, as was shown, for example, by Erikson and Low [39]. We have shown in Fig. 5c the theoretical variation of σ_T as a function of T for the case where the mechanism of de-

---

* Forced elastic deformation is the quasiplastic deformation in high molecular substances. It differs from ordinary plastic deformation in that it is reversible under the effect of temperature.

formation of the crystal changes from slipping to twinning with decreasing temperature (taken from [39]). Figure 5d shows the corresponding experimental curve on which the mechanisms of plastic deformation at different temperatures are indicated.

Garber et al.[40] have investigated the effect of temperature on the flow limit of beryllium during twinning and slipping, showing that the flow limit depends much more on temperature during slipping than during twinning (Fig. 7). He also shows that the character of the temperature dependence of the flow limit during twinning depends on the crystallographic orientation of the planes and the directions of twinning.

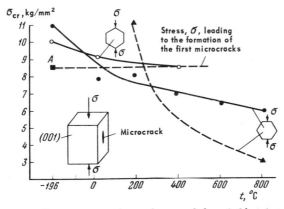

Fig. 7. Temperature dependence of the yield point during twinning of beryllium along the (102) plane for two orientations (solid lines) and during slipping along the (100) and (101) planes (dotted lines) (according to Garber et al. [40]).

The curve in Fig. 5c, analogous to the well-known Ioffe diagram, characterizes the transition from plastic deformation to rupture. According to this diagram the point of intersection of the curve representing the variation of the resistance limit with temperature and the curve representing the variation of the flow limit with temperature determines the conditions under which low temperature embrittlement occurs. As we have said, a special review by Davidenkov and Chuchman [30] is devoted to this problem, and therefore we shall not deal with it here in any more detail.

For the clarification of the slip mechanism in crystals and the theory of plasticity, the fact that σ_T depends very little on T and v is very important. This low dependence, characteristic of hexagonal close-packed and face-centered cubic crystals, has been noted by many authors [18]. Of particular importance is the fact that the low dependence of σ_T on T and v is observed even at temperatures up to the melting point. Many authors believe that this low dependence of σ_T on T when v is constant can be explained by the dependence of the elastic modulus on T. This must mean that the process determining the conditions of the beginning of slip are athermic.

According to [18], the fact that the flow limit is independent of temperature and deformation rate in crystals differentiates crystals from amorphous bodies. However, comparison of the data concerning the variation of. σ_T as a function of T and v for amorphous and crystalline bodies can lead to another interpretation of the difference between the two. This interpretation depends on finding some common factor in the mechanical properties of solids with different structures. This interpretation stresses the common thermodynamic relationships determining the behavior of bodies during deformation, common regrouping mechanisms, and certain common temperatures and time dependences of the parameters of strength and plasticity of bodies with different structures, etc. From this point of view the small dependence of σ_T on T under given conditions does not exclude a large dependence of σ_T on T under other conditions (at deformation rates considerably lower than those indicated in the experiments of Schmid and Boas). Regel' and Govorkov [41] showed that the ratio of the flow limit of zinc monocrystals at $-90°C$ to the flow limit at $+400°C$ increases from 4 to 50 when the deformation rate decreases 10^5 times. To explain such a great effect of T on σ_T it is obviously insufficient to take into account only the effect of temperature on the elasticity moduli: it is necessary to consider that different types of individual and collective atomic regroupings participate in the process of plastic deformation.

If the development of plastic deformation depends on the probability of atomic regrouping then there is no sense in introducing the concept of flow limit or critical shearing stress without taking into account the corresponding time factor. The assumption of the necessity of taking the time factor into account in determining the conditions of the beginning of plastic deformation is analogous to the necessity of taking the time factor into account in determining the resistance limit, as was suggested by Zhurkov [7-10]. By analogy one can expect that with increasing temperature the time dependence will become relatively more important than at low temperature. From this viewpoint the investigations of the effect of temperature of σ_T at temperatures close to the melting point have a particular interest. Unfortunately, after the publication of the monograph by Schmid and Boas there were for practical purposes no experimental investigations of the effect of T and v on σ_T in monocrystals at temperatures close to melting point. Nevertheless, there are indications that in this temperature region the time dependence of the flow limit does become relatively larger than at low temperatures. Such results were obtained, for example, in the case of zinc monocrystals by Regel' and

Govorkov [41]. The wide scattering of the results from this investigation allowed only qualitative conclusions. For practical purposes there are no quantivative results on the time dependence of σ_T at temperatures close to the melting point.

Although the conclusions of Regel' and Govorkov are only qualitative, they contradict the conclusion by Schmid and Boas that at high temperatures the effect of T and v on σ_T and hardening becomes vanishingly small. New experimental data are needed on the effect of T and v on σ_T and hardening for different crystals in order to make a critical analysis of the current opinion.

So far we have discussed only investigations in which was considered the effect of T and v on only one parameter of the elongation curves, namely, the flow limit.

Investigations of the effect of T and v on other parameters of the elongation curve would make it possible to study the kinetics of the process of hardening and annealing. According to Schmid and Boas [18] the competition between these two processes determines the kinetics of plastic deformation beyond the flow limit.

The idea that hardening and annealing are the basic phenomena of plastic deformation was developed by many authors after Schmid and Boas' publication. Kuznetsov [42], Bol'shanina [43, 44], and Vasil'ev et al. [45-49], all explain the temperature and time dependence of the mechanical properties of materials on the basis of the theory of hardening and annealing. In these investigations attempts were made to pass from a qualitative description of the phenomena of hardening and annealing to a quantitative description. In these investigations the equations proposed to describe the kinetics of deformation were checked mainly on the basis of experiments with polycrystalline metals, and therefore we shall not dwell on them any further here. We can note only that all these works [42-49, 18] are based on the assumption that the process of hardening is independent of T and v while annealing depends on them to a considerable degree.

During recent years there has been a relatively small number of publications in which the kinetics of hardening and annealing of monocrystals is characterized by the relationship between the parameters of the elongation curves and T and v; most of these investigations are devoted to face-centered crystals. In this article we have grouped all these investigations in a special section which is followed by a discussion of the results obtained from the viewpoint of the dislocation theory.

There have also been few publications in which the kinetics of relaxation of hardened monocrystals is calculated by determining the variation of the parameter of the elongation curve, as was done by Regel' and Dubov [50], working with TlBr−TlI monocrystals.

The small number of such investigations can be partly explained by the fact that in recent years the kinetics of plastic deformation is more often characterized by the parameters of the creep curve than parameters of compression and elongation curves.

There exist a great number of theoretical and experimental investigations of the creep of mono- and polycrystalline substances. A special review should be devoted to them. However, one of the parameters of the creep curve, namely the rate of steady creep, is apparently directly connected to the parameters of the elongation curves. In this respect the work of Weertman [51] is of great interest. He showed that if $dv/dt = 0$ and $d\sigma/dt = 0$ the functional relationship between the deformation rate, stress, and temperature does not depend on whether these magnitudes were obtained from the curve of steady creep or from the elongation curve at a constant deformation rate. He found that the maximum stress σ_B reached at a given deformation rate corresponds to constant creep stress on the elongation curves apparently in this case, as well, $\left.\dfrac{d\sigma}{dt}\right|_{\sigma = \sigma_B} = 0.$

Thus the stress corresponding to steady flow is independent of the way in which the state of the stable flow is reached. To prove this point Weertman describes experimental results obtained from the creep curves and elongation curves of mono- and polycrystalline aluminum.

As we have previously indicated, Zhurkov and Sanfirova [11] have determined experimentally the dependence of the rate of steady creep on T and σ:

$$v = A \exp\left(-\frac{U_0 - a\sigma}{kT}\right). \tag{9}$$

In this expression the activation energy U_0 is equal to the binding energy and not to the self-diffusion energy.

This result is not in agreement with the empirical and theoretical formulae proposed by other authors. Thus, for example, in his publication on creep, Dorn [12] gives the following expression for v:

$$v = A \cdot F(\sigma) \cdot \exp\left(-\frac{Q}{kT}\right), \tag{10}$$

where $F(\sigma)$ is not an exponential function but a power function; the activation energy Q, as in Weertman's formula, is equal to the activation energy of self-diffusion.

Feltham [52, 53] has investigated high-temperature creep in metals on the basis of the formula proposed earlier by Kauzmann [14]:

$$v = A \exp\left(-\frac{Q}{kT}\right) \text{sh}\left(\frac{a\sigma}{kT}\right). \tag{11}$$

This formula is identical to that of Zhurkov if one replaces the hyperbolic sign by an exponent. However, Feltham shows that the coefficient a, which has the dimensions of the volume, is very much dependent on temperature:

$$a = a_0 \exp\left[\frac{H}{kT}\left(1 - \frac{T}{T_f}\right)\right], \qquad (12)$$

where H is an energy approximately equal to twice the heat of fusion, and T_f is the melting point.

This fact disagrees with the results of Zhurkov and Sanfirova, from which it follows that coefficient a is independent of temperature.

On the basis of the dislocation theory of plasticity of crystals Weertman [54-56] derived the following formula for the steady creep rate:

$$v = A \cdot \sigma^a \cdot \mathrm{sh}\left(B \cdot \sigma^{\frac{b}{kT}}\right) \exp\left(-\frac{Q}{kT}\right) \qquad (13)$$

and considers that it is confirmed experimentally.

Regel', Govorkov, and Dobrzhanskii [57] tried to check the validity of formula (13) by determining the dependence of the limit stress on T and v in experimental studies of the elongation of silver chloride monocrystals. In these experiments the deformation rate varied by five orders at temperatures from −190 to +400°C. The variation of σ_B with v is shown in logarithmic coordinates in Fig. 8. The figure shows that the variation of log σ_B is a linear function of log v but the variation of the slopes of the straight lines, log σ_B = f (log v), with temperature does not correspond to those given by Weertman's formula.

The lack of correspondence between the experimental data and the theoretical formulas in

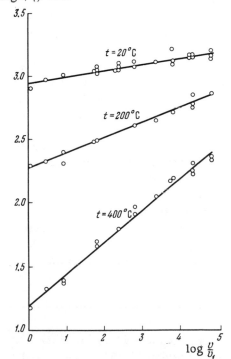

log σ, g/mm²

Fig. 8. Effect of the deformation rate on the steady flow stress for AgCl monocrystals (according to Regel' et al. [57]). v_1) Minimum deformation rate.

this investigation shows the necessity of further experimental and theoretical investigations of this question.

To conclude this part of our review, we must mention the work of Gilman [58], who recently investigated the temperature and time dependence of the plasticity parameters of zinc monocrystals. Gilman investigated slip along the basal and prismatic planes of zinc at temperatures from 250 to 400°C and showed that prismatic slip follows the law of shear stress and the equation of state, which correlate the deformation rate v with the shear stress τ and temperature T in the following way:

$$v = A \cdot \tau^n \exp\left(-\frac{Q}{RT}\right). \tag{14}$$

This constants n and Q in this equation depend on the concentration of impurities. For pure zinc (99.999%) n = 3, Q = 38,000 cal/mole, and A = 700. With increased concentrations of impurities, n and Q increase. Figure 9 gives curves from Gilman's article which confirm the correctness of Eq. (14). Figure 9a indicates that at T = const. the flow limit is a linear function of the deformation rate in logarithmic coordinates. Figure 9b illustrates the linear relationship of log v as a function of 1/T with τ = const., and Fig. 9c shows the linear relationship between log τ and 1/T for v = const. Figure 9d shows that the addition of 0.1% Cd displaces the log τ = f (log v) curves and changes their slope. Figure 9a, b, c, and d are relative to prismatic slip. Figure 9a represents the variation of critical shear stress in the case of slip along the basal plane in zinc crystals containing 0.1% Cd as a function of temperature for two deformation rates differing by a factor of 20.

Insofar as within the temperature range of 250-400°C prismatic slip satisfies the equation of state, one can consider that for practical purposes hardening does not occur in this temperature range during prismatic slip.

The results of Gilman's study show that the temperature dependence of slip in a given crystal can be different along different crystallographic planes, which is in agreement with the conclusion by Stepanov and his co-workers [19, 20] cited earlier. The form of the empirical equation given by Gilman is closer to the equation given by Dorn and Weertman than to the Zhurkov equation.

Fig. 9. Effect of temperature T and deformation rate v = dγ/dt on the critical shear stress σ_{cr} for Zn crystals during prismatic (a, b, c, d) and basal (e) slip (according to Gilman [58]). a) Effect of v on τ_{cr} with T = const. in the case of prismatic slip; b) effect of T on v with τ = const. for pure zinc (99.999%) and Zn + 0.1% Cd; c) effect of T on τ_{cr} with v = const; d) effect of v on τ_{cr} at different T for pure Zn (99.999%) and Zn + 0.1% Cd.

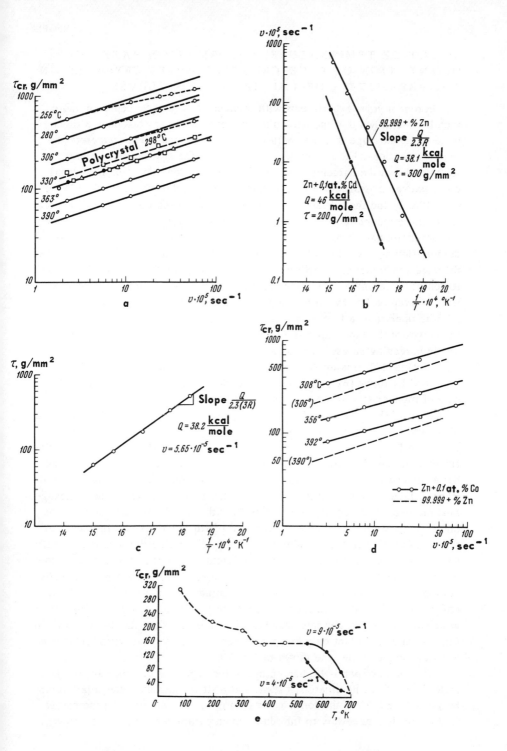

EFFECT OF TEMPERATURE, DEFORMATION RATE, AND ORIENTATION OF FACE-CENTERED CUBIC CRYSTALS ON THE PARAMETERS OF THE HARDENING CURVES

In recent years experiments with monocrystals of face-centered cubic metals have shown that the hardening curves of these crystals have a much more complex shape than was believed earlier and that the parameters of these curves depend to a considerable extent on the orientation. In Fig. 10 we have drawn the hardening curves of face-centered cubic and hexagonal close-packed crystals according to data given by Schmid and Boas [18] (Fig. 10a, b) and the hardening curves of face-centered cubic crystals according to more recent data (Fig. 10c). The curves of Fig. 10a and b show linear and parabolic hardening, respectively. Linear hardening occurs as a result of deformation of hexagonal close-packed crystals by slip, e.g., Zn and Cd. In this case the hardening coefficient θ is relatively small, and therefore, the region of hardening is called the region of easy glide. Parabolic hardening was observed earlier by Karnop and Zaks in the case of Al monocrystals at room temperature, a fact mentioned by Schmid and Boas [18]. The hardening curves of the type shown in Fig. 10c, whose shape differs from that of parabolic hardening curves, were obtained recently in the case of Al monocrystals (Masing [19], Haasen and Leibfried [60], Lücke and Staubwasser [61], Jaoul and Lacombe [62], Andrade and Henderson [63], etc.) and also in the case of monocrystals of other face-centered cubic metals (Andrade and Henderson [63] in the case of Ag, Au, Ni, Andrade and Aboav [64] in the case of Au and Cu, Diehl [65] and Seeger et al. [66], in the case of Cu, etc.).

From Fig. 10c it can be seen that three stages of deformation can be distinguished. The first stage (I) corresponds to easy glide; here the hardening has a linear character with a low coefficient θ_I. Some authors equate this slip stage in face-centered cubic crystals with the process of easy glide in hexagonal close-packed metals. Apparently in both cases slipping occurs only along one acting system of planes and directions of slip. This is also indicated by the relief of edge surfaces of deformed samples. During the second stage (II) hardening increases linearly with a coefficient θ_{II}, in this case higher by one order of magnitude than in the easy glide stage. Finally, the third stage (III) corresponds to a gradually decreasing hardening rate—the variation of hardening with deformation passes from linear to parabolic. Curves of this type are characterized by a greater number of parameters than the curves shown in Fig. 10a and b. In fact only two parameters, τ_0 and θ_I, are sufficient to characterize curves of the type shown in Fig. 10a.

The curve of parabolic hardening, of the type in Fig. 10b, can be described with three parameters, τ_0, m, and n, if we introduce the relationship between τ and γ, of the form: $\tau = \tau_0 + m\gamma^n$. To characterize the curves in Fig. 10c it is necessary to introduce seven parameters: $\tau_0, \tau_1, \tau_2, \theta_I, \theta_{II}$,

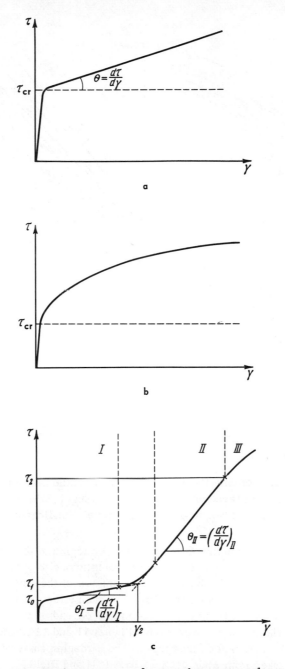

Fig. 10. Hardening curves for metal monocrystals
a) Hexagonal close-packed crystals (according to
Schmid and Boas [18]); b) face-centered cubic
crystals (according to Schmid and Boas [18]); c)
face-centered cubic crystals (according to Diehl [65]).

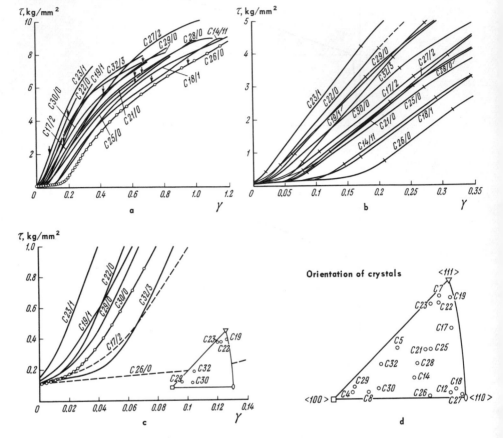

Fig. 11. Effect of orientation on the parameters of the hardening curves of copper monocrystals (according to Diehl [65]) a and b; the curves are drawn to different scales; the orientation of the samples is indicated in c and d.

and two more to characterize the third parabolic region of the curve. It is clear that in this case it is infinitely more complicated to obtain experimental data on the effect of T and v on the parameters of the curve. The experimental investigation is particularly difficult because the parameters of the hardening curves of face-centered cubic metals depend on the original crystallographic orientation of the samples. Figures 11 and 12 illustrate the effect of crystal orientation on the shape of the hardening curve of Cu and Ag monocrystals taken from Andrade and Aboav [64] and Diehl [65]. Diehl [65] analyzed these curves and showed that all these hardening curves can be divided into two groups in terms of the effect of the orientation on the parameters of the curve: the parameters τ_0, τ_1, θ_{II}, and the extent of the stage of

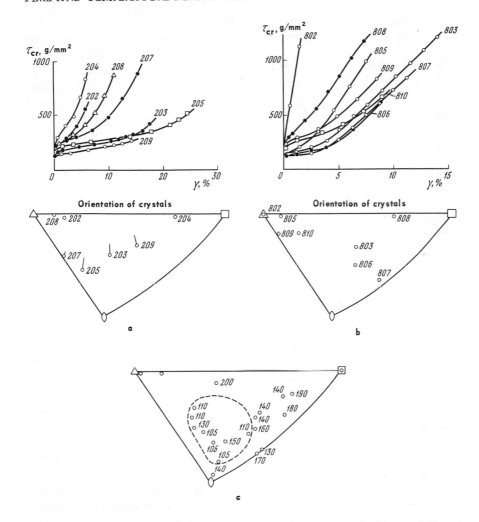

Fig. 12. Effect of orientation on the parameters of the hardening curves for copper (a) and gold (b) monocrystals (according to Andrade and Aboav [64]); the orientations of the samples are indicated in the figures; the value of the critical shear stress for different orientations is indicated in c.

easy glide γ_2, and depending very little on orientation, while the dependence of θ_I is somewhat greater. The difference between the hardening curves relative to different orientations depends basically on the difference of the slope in the initial part while the dependence of γ_2 on orientation has a secondary importance due to the small dependence of τ_0 and τ_1 on orientation.

The temperature dependence of all these parameters has been studied very little and the existing results have a qualitative character. Figure 13 represents the theoretical variation of the shape of the hardening curve with temperature. A few experimental curves obtained at different temperatures with Cu and Ag monocrystals by Andrade and Aboav [64] and Diehl [65] are given in Fig. 14. According to Diehl, analysis of the experimental results shows that the slope θ_{II} in the second stage depends very little on the temperature, at least for Cu in the temperature range of 4.2-300°K. The origin of stage III is displaced toward larger values of τ with decreasing T. For copper the temperature dependence of stage I is expressed by the decrease of its extent with decreasing T. The results of the investigation of Ni, Ag, and Au coincide in general with the results obtained by Diehl for Cu. In the case of Ni, Ag, and Au there is a linear stage II, with decreasing temperature. However, for these metals as well as for copper (according to [64]), stage I decreases with increasing temperature; (compare Fig. 15 with Fig. 14a, b). Diehl explains the variation of θ_{II} observed by some investigators not by the effect of temperature but by the effect of the orientation of different samples, which was not taken into account by some authors.

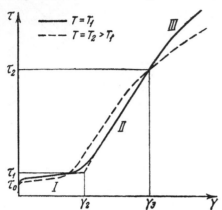

Fig. 13. Variation of the shape of the hardening curves of copper with temperature (schematic) (according to Seeger et al. [66]).

Monocrystals of technically pure Al, which are often taken as standard face-centered cubic metals, are an exception. At room temperature region II is absent in the case of Al. Diehl considers that this result is possibly due to the effect of impurities in the aluminum.

Contradictions in the results of the investigations of different authors indicate the complexity of the experiments meant to determine the effect of orientation, temperature, and deformation rate on the parameters of the hardening curves of face-centered cubic crystals. We shall return to this problem later in connection with the investigation by Seeger et al. [66].

The purpose of a number of recent investigations of the effect of temperature on the parameters of the deformation curve consists in separating the reversible part σ_r of the total stress σ, necessary to achieve a given deformation ε, from the irreversible σ_{irr}. The reversible part depends on the tem-

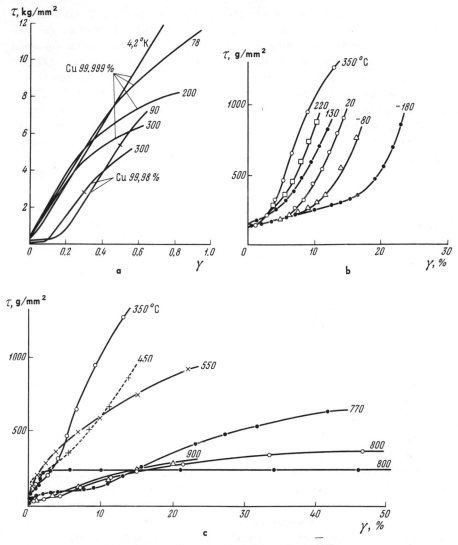

Fig. 14. Hardening curves of copper monocrystals at different temperatures. a) According to Diehl [65]; b, c) according to Andrade and Aboav [63].

perature of the experiment. Earlier, such investigations were carried out with polycrystals. We shall consider in more detail two investigations of Al and Cu monocrystals by Cottrell et al. [67, 68].

If two metal samples are deformed at an identical rate but at different temperatures their deformation curves usually diverge with increasing temperatures. To calculate this divergence we shall introduce the ratio AB/AD at a given ε (Fig. 16). Cottrell and Stokes [67] consider the difference in σ at different temperatures to be the result of two processes. First of all the amount of hardening induced by a given deformation may depend on the deformation temperature. The hardened state in B can differ from that in C. Secondly, for a given hardened state σ can vary reversibly with the temperature. The method of separating these two effects, according to Fig. 16, consists in deformation of the sample up to point B at temperatures T_2 and then continuing the deformation at T_1, which results in the curve ABD. If between the measurements B and D aging was eliminated and there was no annealing, then BD and DC are respectively the measures of the relative variation of stress flow and of the change in the level of hardening reached at two different temperatures.

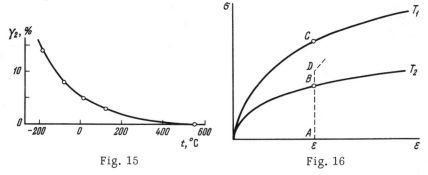

Fig. 15

Fig. 16

Fig. 15. Effect of temperature on the degree of deformation of easy glide (ε_2 in Fig. 12) (according to Andrade and Aboav [64]).

Fig. 16. Deformation curves (schematic) at two temperatures, $T_2 > T_1$ (according to Cottrell and Stokes [67]).

In principle the ratio AB/AD can be determined as a function of T and ε by deforming several identical samples up to different ε. This is possible in the case of polycrystalline samples but impossible in the case of monocrystals because of the individual variations in their plastic properties. The necessary data can be obtained, however, by studying a single sample since the ratio sought, AB/AD, remains constant when the temperature changes from T_2 to T_1 if the deformation is increased by small steps. The ratio AB/AD is equally independent of the ε at which it is measured or the temperature or mechanical history of the sample. Cottrell and Stokes indicate that in the case of Al this ratio is also independent of the orientation.

The results of measurements of the effect of temperature on the ratio AB/AD for Al monocrystals are shown in Fig. 17b. The dependence of AB/AD on temperature if partially related to the dependence of the elastic modulus, σ/E, on temperature. Therefore in Fig. 17a we have drawn two curves:

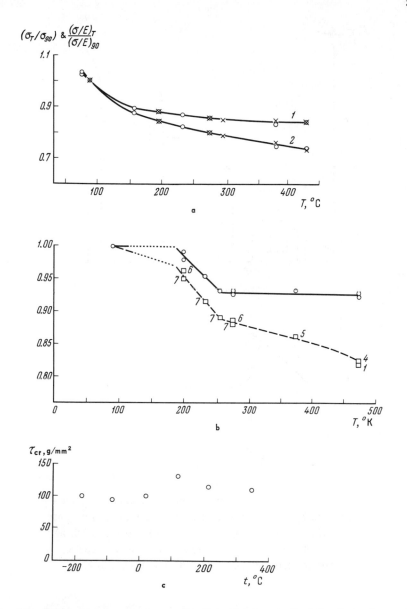

Fig. 17. Effect of temperature on the flow stress. a) For aluminum (according to Cottrell and Stokes [67]): Curve 1) direct measurements; Curve 2) measurements taking into account the variation of elastic moduli with temperature; b) for copper (according to Adams and Cottrell [68]): the upper curve is drawn by taking into account the variation of the elastic modulus with temperature; c) for copper (according to Andrade and Aboav [63]).

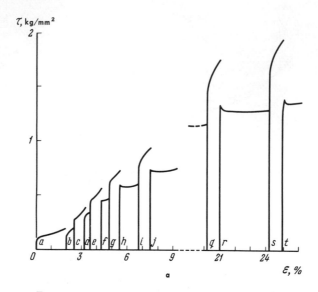

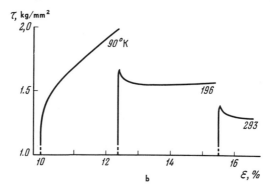

Fig. 18. Curves for determination of the annealing phenomenon by deformation of the same crystal at alternate
temperatures (according to Cottrell and Stokes [67]). a)
Curves obtained by alternating between 90 and 293°K; b)
the sample tested at 90, then 196, and then 293°K.

curve I represents the variation of the ratio AB/AD as a function of T, and
curve II the variation of σ/E as a function of T. It can be seen that in the
range of temperature from 200 to 400°K, curve II is almost horizontal. This
means that the flow stress (a magnitude analogous to limit flow) is completely independent of temperature and deformation rate, at least in this temperature range.

The method of alternate variation of temperature to obtain deformation curves analogous to those shown in Fig. 18 was used by Cottrell and Stokes to characterize the phenomenon of annealing. The visible peaks on curves obtained at higher temperatures after preliminary testing at low temperatures are explained by the annealing process which develops during deformation under the effect of stress. The authors showed that for practical purposes the effect of aging is absent in this case. The appearance of the peaks is partially explained by geometric annealing due to the rotation of the plane and the direction of slip during deformation of local areas of the crystal. According to the calculations of these authors, however, geometric annealing may account for only one-eighth of the total loss of stress corresponding to the peak. The process of annealing, which is indicated by the appearance of the peak, is connected to the formation of Luder's lines. The effect of stress on the annealing process consists in the decrease of the activation energy necessary to overcome the disarrangements of the structure resulting from hardening.

Adams and Cottrell [68] investigated the effect of temperature on the parameters of the elongation curves of Cu monocrystals by the same method of alternate variation of temperature used by Cottrell and Stokes. The curve representing the variation of the flow stress as a function of temperature obtained in this investigation is given in Fig. 17b. The introduction of corrections accounting for the temperature dependence of the elastic modulus showed the presence of a horizontal section of the curve, $\sigma = f(T)$, as was seen in Fig. 17a, which indicates that the slip process in Cu monocrystals is not dependent on temperature or deformation rate. For the sake of comparison we have shown in Fig. 17c the data obtained by Andrade and Aboav [64] on the effect of temperature on the flow limit in Cu.

The discussion of the shape of the hardening curves of face-centered cubic metals and the temperature dependence of the parameters of these curves from the viewpoint of the dislocation theory was based on the results of the investigations cited here. This discussion is given in detail by Seeger and other authors. Later we shall present their basic conclusions, but first we shall describe some concepts of the dislocation theory of plasticity given by Schoeck [69] which concern the flow limit and hardening of crystals. For brevity we shall omit most of the literature cited by Schoeck [69].

DISCUSSION OF THE DATA ON THE TEMPERATURE DEPENDENCE OF THE FLOW LIMIT FROM THE VIEWPOINT OF THE DISLOCATION THEORY OF PLASTICITY

According to the dislocation theory, flow in crystals means that a considerable number of dislocations move through the crystal lattice over large distances. Critical stresses, which determine the beginning of slip, are explained by the fact that either the generation of dislocations or their motion requires the application of definite critical forces. At the present time there are several hypotheses of the conditions necessary for intense slip to begin. Briefly, these hypotheses can be reduced to the following:

a) Frank-Read Sources. The critical shear stress is determined by the stress of the Frank-Read sources at which they begin to generate dis-

locations. If the length of the source is l, the stress at the source is $\tau_i = \dfrac{G \cdot b}{l}$,

where G is the shear modulus and b is Burgers' vector. The beginning of slip will be sharply defined if the length of the sources present in the crystal change very little with respect to a certain average length, l_{av}.

b) Interaction of Dislocations at Large Distances. The conditions of flow are determined by the interaction between dislocations. For example, two parallel edge dislocations in parallel slip planes whose distance is r can pass each other going in opposite directions if the ex-

ternal stresses reach a critical value $\tau_i = \dfrac{G}{8\pi\,(1-\nu)}\dfrac{b}{r}$, where ν is the Poisson

coefficient. Since not all dislocations interact, the distance r must be somewhat greater than the average distance r_0 between dislocations. If ρ is the density of dislocations, then r is of the order of $\rho^{-1/2}$

c) Peierls Forces. Even in the ideal lattice some force is necessary for the motion of dislocations. To calculate these forces one must conceive of some distribution of the atoms in the dislocation nucleus. Peierls' forces are calculated on the basis of Peierls' dislocation model. The resulting critical stress depends to a large extent on the atomic structure of the central part of the dislocation. Since the flow limit depends on the microstructure and the purity of the material, and Peierls' forces must be independent of them, it is usually assumed that in typical metals Peierls forces are below the flow limit. However, there are indications that in crystals with strong bonds, in diamonds, for example, Peierls forces can reach such high values that it is these forces which determine the flow limit. It should be noted that Indenbom [70] came to the same conclusion as the result of an investigation of the Frenkel'-Kontorova slip model.

d) Local Disarrangements of Structure. When a dislocation moves in a slip plane it must overcome certain local disarrangements of structure. Such disarrangements may be other dislocations crossing the slip plane, groups of precipitated particles, or imbedded atoms. The dislocation must have a sufficiently high energy to overcome these barriers. Even if the thermal fluctuations help to overcome these barriers, part of the energy must be spent for the motion of dislocations because of the work due to external forces. Thus we come again to the critical stress necessary for the motion of lines of dislocation through the lattice with a given velocity. The value of the stress depends to a large extent on the initial distribution of the disarrangements of structure.

e) Pinning Down of Dislocations due to Elastic Interaction with Dissolved Atoms. The segregation of impurity atoms near the dislocation due to elastic interaction can pin down a dislocation in a given position. This explains, for example, the peak of flow observed in α-iron which contains a small amount of carbon or nitrogen. This type of pinning down of dislocations occurs specifically in body-centered cubic crystals, where the impurities induce disarrangements of structure which interact with edge dislocations as well as screw dislocations. Only a small amount of impurities is necessary to pin down the dislocations. Usually 0.01% or even less forms a braking atmosphere.

Many investigators consider that the flow limit τ_0 in pure metals is determined by the length l of the Frank-Read source. The upper limit l can be calculated from the experimental value of τ_0. For Al and Cu, $l = 10^{-3}$ cm and for Zn, $l = 10^{-5}$ cm. Seeger [71] criticizes this point of view and indicates that it is difficult to understand why a distribution function along the length of the source has such sharp limits, which are necessary to explain the sudden appearance of plastic flow.

Edwards, Washburn, and Parker [72] have interpreted some experiments by Edwards, coming to the conclusion that the flow limit is determined by the field of internal stresses τ_i of the dislocation net. According to these experiments, the stresses necessary for the motion of dislocations forming the boundary of blocks slightly rotated (with respect to each other) do not differ from the yield point. It was also shown that a monocrystal with a high dislocation density has a high yield point. For a reasonable amount of dislocation density ($\rho \approx 10^7$ cm^{-2}) internal stress in an undeformed crystal is of the order of magnitude of the observed yield point. The initial stress of a part of the Frank-Read source in this case is less than τ_i.

Which of the two mechanisms a or b determines the beginning of the flow depends on the initial structure of the dislocation net. At the present time there is only qualitative information on the structure, obtained by x-ray investigations and measurements of internal friction. The etching and decoration methods, which put the dislocations in evidence, are very promising but so far have given only preliminary results.

In cases a and b the flow limit must depend very little on temperature. The regions of the crystal in which dislocation must be activated are so large that thermal fluctuations cannot occur. The temperature dependence must be of the same type as that of the shear modulus. For given metals within given temperature ranges the temperature dependence of the flow limit is greater than one would expect for the temperature dependence of the shear modulus. Therefore there must be another mechanism determining the flow.

Among the mechanisms mentioned, the mechanisms by which the motion of dislocations is opposed by Cottrell impurities or local disarrangements of structure are sensitive to temperature. In these cases the activation energy must be transferred to small regions, and therefore thermal fluctuations can substantially increase the effect of external stresses on the motion of dislocations.

The Cottrell mechanism of the pinning down of dislocations by the atmosphere of impurity atoms was used to explain the considerable temperature dependence of the yield point in body-centered cubic metals (Cottrell and Bilby [73]). As a result of thermal fluctuations a small dislocation loop can detach itself from the atmosphere which holds it. The acting external forces tending to enlarge the loop and move it forward can exceed the slowing down forces. The loop will increase and finally the whole line of dislocations will break off from the atmosphere which retains it. We have shown earlier (Fig. 5b) that the calculated temperature dependence of the flow in α-iron containing 0.003% C is in agreement with the experimental data.

Fisher [74] showed theoretically that in the case of the Cottrell mechanism the yield point σ_T at high temperatures is inversely proportional to the temperature, i.e., $\sigma_T \cdot T = $ const. This relationship is in agreement with the experimental data concerning Mo and Fe. At lower temperatures the experimental values of σ_T are below the theoretical curve $\sigma_T \sim 1/T$, and at sufficiently low temperatures reach critical values corresponding to the stress necessary to break off dislocations from the atmosphere of impurities without thermal activation.

Erikson and Low [39] showed that for α-iron the temperature dependence of the yield point at very low temperatures is actually somewhat different from the represented in Fig. 5a, b. Near G = 0 the curve has a section parallel to the abcissa (Fig. 5d). We have mentioned before, however, that Erikson and Low [39] explained this by the transition from slipping to twinning without referring to the Fisher theory.

The results of the investigation by Pearson, Read, and Feldman [75] showed that for very thin hairlike crystals of silicon, whose strength approaches the theoretical value, the yield point depends very much on the temperature (Fig. 19). Thus the deformation curves have very sharp yield points similar to those occurring in α-iron containing carbon. The authors note that the temperature dependence of the yield point and the phenomenon of aging in silicon can be explained on the basis of the Cottrell mechanism, taking into account the fact that silicon can contain up to 0.01% oxygen atoms. This amount of impurities is sufficient to explain the effects observed. However, the quantitative dependence of σ_T on T in this case is sharper than predicted by Fisher on the basis of the Cottrell mechanism. In the temperature range of 600 to 800°C σ_T is proportional to $1/T^5$ and not to $1/T$.

The authors indicate that in the case of silicon, as for other semiconductors, the presence of covalent bonds is characteristic and the temperature dependence of the yield point and the phenomenon of aging can be due to Peierls forces (Dietze [76]). In the equilibrium state dislocations are located in the directions corresponding to minimum energy but the motion of such dislocations is hindered by large Peierls forces. Due to the diffusion process, dislocations can become reoriented in such a way that Peierls forces will be smaller although the energy of such dislocations will be larger. The process of aging consists of dislocations returning to the state of lower energy but higher Peierls forces.

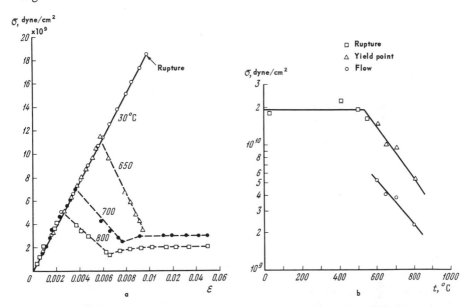

Fig. 19. Deformation curves of threadlike silicon crystals at different temperatures (a) and the variation of the resistance limit and of the upper and lower yield points of these crystals as a function of temperature (b) (according to Pearson, Read, and Feldman [75]).

Seeger [71, 77, 78, 79] assumes that in face-centered cubic and hexagonal close-packed materials the increase of the yield point at low temperatures is due to the effect of local disarrangements of structure and particularly to the intersection of dislocations. In close-packed structures complete dislocations are usually transformed into extended dislocations composed of partial dislocations limiting the stacking fault. The activation energy of the intersection of extended dislocations is particularly high, since the partial dislocations must recombine at the intersection point in order to avoid a consid-

erable change of atomic structure. Thus, according to Seeger the temperature dependence of the yield point of close-packed crystals must be similar
to those represented in Fig. 20a, b. At low temperatures the intersection of
dislocations, formation of jogs, or formation of trails of vacancies or interstitial atoms behind the moving jogs are the effects determining the linear dependence of σ_T on T (line A, Fig. 20a). If the temperature is above a given
value, T_0, all these processes occur so easily that the flow is determined only
by internal stresses, which depend very little on temperature (line B, Fig. 20a).
If the activation energy necessary for the intersection of dislocation lines is
high then, according to Cottrell, at low temperatures it is more advantageous

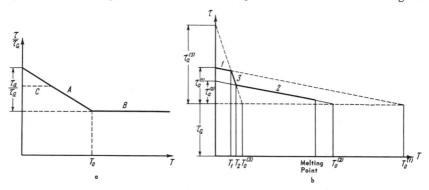

Fig. 20. Schematic representation of the temperature dependence of the flow
stress calculated by Eq. (16) a and for copper b (according to Seeger et al. [79]).

from the energy standpoint for the dislocation to "pass" around the forest of
dislocations than to cross it. A moving dislocation partly squeezes between
the dislocations crossing the slip plane and partly goes around, recombining
on the opposite side. This mechanism must be independent of temperature
because the effect of thermal activation is absent (line C, Fig. 20a). Such an
effect of temperature on the yield point is observed in the case of hexagonal
close-packed materials. For face-centered cubic crystals Seeger [80] has distinguished two cases, depending on whether the distance between the two half
dislocations is large (gold and copper) or small (aluminum). In the first case
the temperature dependence, which is connected with the intersection mechanism, is very small (slope of line A, Fig. 20a); in the second case it is very
large.

   In some cases the dislocation theory attempts to give quantitative expla-
nations of experimental facts. It was mentioned before that Cottrell and
Bilby [73] have developed a quantitative theory of the effect of temperature
on the yield point of the body-centered cubic crystals and the experimental
data were in satisfactory agreement with the theory. Seeger [79] has ex-

plained the effect of temperature on the flow stress in face-centered cubic crystals not only on the basis of the qualitative conclusions mentioned above, but also on the basis of some quantitative relationships. He considers that in face-centered cubic and hexagonal close-packed crystals the temperature dependence of the stress necessary for their deformation at temperatures below the self-diffusion temperature ($T < T_0$ in Fig. 20) is determined essentially by the intersection of dislocations and formation of lattice defects resulting from jogs. On the basis of this assumption he indicates that the energy of activation for these processes is a linear function of the shearing stresses acting on the dislocation, since during activation the dislocation moves forward and the work performed by the effective stress τ_s acting on the dislocation is equal to $\tau_s \cdot d$, where d is the distance at which activation occurs.

The effective stress τ_s is equal to the difference between the applied external stress τ and the stress impeding the motion of dislocation τ_G. The stress τ_G depends only indirectly on the temperature through the shear modulus.

If only processes induced by thermal activation occur then the flow rate, according to Seeger, can be expressed by the following formula:

$$\dot{\varepsilon} = N \cdot A \cdot b \cdot \nu_0 \exp\left\{-\frac{F_0 - v(\tau - \tau_G)}{kT}\right\}. \tag{15}$$

Here N is the number of places in which activation acts per unit volume occur; A is the area covered by a moving line of dislocations as a result of each activation act; ν_0 is the oscillation frequency, of the order of the Debye frequency; F_0 is the free activation energy in the absence of stress; and v is the volume of activation. For a dislocation with a Burgers vector b, the activation volume $v = b \cdot d \cdot l$, where l is the distance between places where the lines of dislocations were impeded. The activation distance d is approximately equal to the interatomic distance b in processes in which vacancies or interstitial atoms occur and also in intersections of extended dislocations. When extended dislocations intersect d depends on the width of the dislocation which is intersected and on its orientation with respect to the intersecting dislocation.

The solution of Eq. (15) in terms of stress as a function of the deformation rate ε and temperature T gives a linear relationship between the ratio τ/τ_G and temperature within the temperature range of 0 to T_0:

$$\left.\begin{array}{l}\tau = \tau_G + \dfrac{H_0 - T\left\{S + k\ln N \cdot A \cdot b \cdot \dfrac{\nu_0}{\dot{\varepsilon}}\right\}}{v} \quad \text{at } T < T_0, \\[4mm] \tau = \tau_G \quad \text{at} \quad T > T_0,\end{array}\right\} \tag{16}$$

where

$$T_0 = \frac{H_0}{S + k\ln\left(N \cdot A \cdot b \cdot \dfrac{\nu_0}{\dot{\varepsilon}}\right)}. \tag{17}$$

In these expressions the free activation energy F_0 is given in the form

$$F_0 = H_0 - T \cdot S,$$

where S is the activation entropy.

This type of temperature dependence of τ is shown in Fig. 20a.

Seeger made a detailed investigation of the temperature dependence of τ_s in the case of copper, assuming that it is the result of three possible processes: 1) the intersection of an edge dislocation with a forest of dislocations; 2) the intersection of a screw dislocation with a forest of dislocations; 3) jogs of screw dislocations, with formation of vacancies. In the case of copper one can calculate the activation energy for these three processes, and for the third process the activation entropy can also be calculated. The variation of τ as a function of temperature for copper is drawn on the basis of these calculations (Fig. 20b) and this curve coincides with the experimental data of Adams and Cottrell [68] (Fig. 17b). The temperatures at two intersection points turn out to be $T_1 = 180°K$ and $T_2 = 250°K$. Outside this temperature range the temperature dependence of τ must be very small compared with that of the shear modulus, according to Seeger. Therefore it is very difficult to separate the temperature dependence of τ due to the thermally activated intersection of dislocations from that due to the temperature dependence of elastic constants and the energy of stacking faults. As a result it is easier to check the theory by studying the effect of the change of the deformation rate on τ instead of the effect of temperature.

We can note here that the formula obtained by Seeger for the flow rate resembles the formula derived by Zhurkov [9]. However, Feltham [52, 53] indicates that there is a certain difference between the two formulas [see formula (12)].

We consider it necessary to describe in more detail the results obtained by Seeger on the effect of temperature on the yield point because this is one of his most recent works.

DISCUSSION OF THE ORIENTATION AND TEMPERATURE
DEPENDENCE OF THE PARAMETERS OF HARDENING
CURVES OF FACE-CENTERED CUBIC METALS FROM THE
VIEWPOINT OF THE DISLOCATION THEORY

The theory of the occurrence and development of intense slip were based on the concept of dislocation. The theory must explain the phenomenon of work hardening, the difference in the shapes of work hardening curves of crystals with different structures, and the effect of crystallographic orientation, temperature, and deformation rate on a number of parameters of these curves.

The first attempt to explain work hardening on the basis of dislocation was made by Taylor [81]. He assumed that the density of dislocations in the crystal— and consequently their interactions— increase during deformation. By introducing a rather artificial scheme of regular and uniform distribution of dislocations in work-hardened crystals, Taylor obtained an equation for a parabolic hardening curve. Mott proposed a more realistic scheme of the distribution of dislocations in work-hardened crystals. The presence of barriers to the motion of dislocations (grain boundaries or sessile dislocations) results in piling up of groups of dislocations at these barriers. It is the interaction between these groups which produces the work hardening of the crystal. The hardening curve calculated on this basis also has a parabolic shape.

The presence of groups of dislocations piled up at barriers results in hardening not only because the stress created by these groups impedes the motion of dislocations in neighboring slip planes but also because these stresses decrease the effect of external stresses in the place where the Frank-Read sources (which induce these stresses) are located. Dislocations created by Frank-Read sources and stuck at a barrier some distance from the source lock the source as the result inverse stresses. However, the stress of this same group of dislocations can induce the generation of a neighboring source. This type of reasoning is used, for example, by Leibfried and Haasen [83] to explain the effect of sharply defined slip plackets on the assumption that the velocity of dislocation is small compared with the velocity of sound, and therefore that the sharply defined slip lines cannot be formed by one Frank-Read source, as Mott assumed. Figure 21 shows a scheme of periodic distribution of Frank-Read

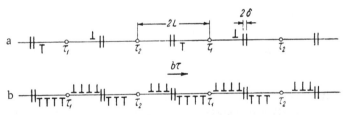

Fig. 21. Model of the distribution of dislocations in a hardened slip plane at the stage of uniform slip a and a model of a hardened plane with edge dislocations of different signs and Frank-Read sources with different starting stresses τ_1 and τ_2, b; the width of the barriers is 2δ (according to Leibfried and Haasen [83]).

sources at distances of 2L with critical yield points τ_1 and τ_2 ($\tau_1 < \tau_2$). Disarrangements of structure with a width of 2δ ($\delta \ll L$) are located between the sources. Under the effect of external stress τ, sources with lower critical stresses τ_1 begin to form dislocations which stick at the disarrangements of structure. The stress field of this group favors the activation of the sources with τ_2 located in the same slip plane. These sources can then form dislocations even when $\tau < \tau_2$. Sharply defined dislocation lines may be formed as the result of the annihilation of dislocations with different signs. This can occur if the concentration of stresses at the disarrangements of structure reaches such a high value that dislocations can pass through them. By this mechanism microscopic slip lines can form during a short period of time even if the velocity of the dislocation is small with respect to the velocity of sound.

It follows then that barriers to the motion of dislocations in the slip plane play an essential role in the dislocation theory of hardening. The nature of these barriers, the conditions of their occurrence and disappearance, and also the possible means by which dislocations overcome them determine the process of work hardening and annealing. These problems are discussed in many investigations on the basis of the shape of the hardening curve and the dependence of the parameters of these curves on T and v, and also on the basis of the investigation of the relief of edge surfaces of deformed samples, of the configuration of the dislocation net revealed by etching and decoration, optical investigations of transparent crystals, x-ray investigations, etc. This problem is taken up in greater detail in the case of close-packed structures.

In face-centered cubic crystals work hardening is explained by many authors by considering the basic barriers to the motion of dislocations to be the so-called Lomer-Cottrell sessile dislocations. They can form at the intersection of two acting slip systems as the result of the encounter of two dislocations, one of which has slipped in the primary system and the other in the secondary system. As a rule complete dislocations in close-packed structures are broken into half dislocations between which a stacking fault occurs. Such a dislocation is called an extended dislocation. Its extension—the distance between the two halves—is the greater the smaller the energy per unit area of the stacking fault. As a result of the interaction of two extended dislocations at the intersection of slip planes of different systems, the dislocations unite into one extended dislocation and pass from one slip plane onto another. A relatively large activation energy is necessary to destroy such Lomer-Cottrell dislocations. The energy of stacking faults in different close-packed crystals, and also the activation energy of Lomer-Cottrell dislocations, can be calculated on the basis of the work of Schoeck and Seeger [84].

In a recent work Seeger et al. [66] describe experiments made to clarify the hardening and annealing mechanism in face-centered cubic crystals and develop a theory in which the Lomer-Cottrell sessile dislocations play an important role.

Seeger [66] investigated the hardening curve of face-centered cubic metals and divided it into three regions, as has been done before by others. In Fig. 12 we have shown typical hardening curves of face-centered cubic metals at two different temperatures, taken from Seeger [66]. In this work the region of easy glide I is not discussed. The investigation of the causes of increased

hardening in region II led the author to the conclusion that it is connected
with the formation of barriers— Lomer-Cottrell sessile dislocations— during deformation. Seeger criticized the view expressed by Friedel [85], who assumed
that the effect of secondary slip systems is necessary only at the beginning of
region II. The increase of hardening in region II requires continuous slip in
the secondary slip system as a result of which new Lomer-Cottrell dislocations
are continuously formed. Seeger also rejects the opinion of Paxton and
Cottrell [86] that the main effect of secondary slip systems is to increase the
density of dislocation forests, which must be crossed by dislocations moving
in the primary slip system. This is confirmed by the low temperature dependence of hardening in region II. The effect of the intersection mechanism
should lead to a much larger temperature dependence.

Experiments and the schematic curves in Fig. 10 indicate that the temperature dependence of hardening in region III is greater than in region II.
The beginning of region III is displaced with temperature, i.e., the value of
τ_2 in Fig. 10 changes. Friedel [85] and Cottrell and Stokes [67] assume that
this is due to the dislocations passing through the barriers along the slip planes.
Analysis of the experimental data concerning the structure of slip bands occurring in stage III showed that the opinion expressed earlier by Diehl is more
correct. According to this view the simultaneous action of stress and thermal
fluctuation in stage III induces transverse slipping, which consists of screw dislocations passing around the barriers by passing from the primary slip plane
onto the transverse plane and then again onto the primary plane.

The Lomer-Cottrell dislocations are formed along the intersection of two
{111} planes, and consequently in the <110> direction. Their position is

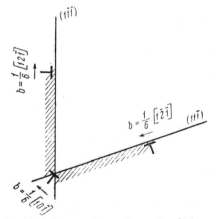

shown in Fig. 22. The band of
stacking faults passes from one octahedral plane onto another. The
combination of partial dislocation
is connected to both planes and
cannot either slip or move away
from the intersection line of these
planes without expenditure of considerable energy.

The Lomer-Cottrell dislocation
shown in Fig. 22, can be formed in
two ways: either by interaction of
dislocations of the $\frac{1}{2}[0\bar{1}\bar{1}]$, $(11\bar{1})$
type with the $\frac{1}{2}[110]$, $(1\bar{1}\bar{1})$ dislocation or by interaction between
dislocations of the $\frac{1}{2}[1\bar{1}0]$, $(11\bar{1})$
and $\frac{1}{2}[0\bar{1}1]$, $(1\bar{1}\bar{1})$ types.

Fig. 22. Lomer-Cottrell sessile dislocations composed of three partial edge
dislocations (according to Seeger et al.
[66]).

The following discussion is based on the assumption that the primary slip system coincides with the $^1/_2$ [101], (11$\bar{1}$) system. In this octahedral system shear stresses are maxima if the direction of elongation is situated in the middle of the stereographic triangle, indicated by the square hatching in Fig. 23.

Each triangle in Fig. 23 represents a different slip system in which the shear stress is maximum if the axis of elongation is in the center of the triangle.

The direction of Lomer-Cottrell dislocations obtained by interaction with a given system is indicated by the respective directions of the hatching lines (solid or dashed).

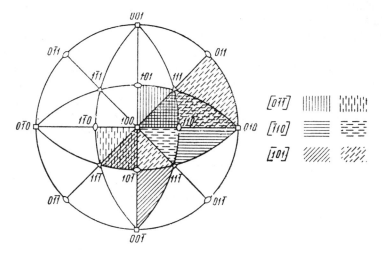

Fig. 23. Combinations of slip systems capable of forming Lomer-Cottrell sessile dislocations during flow. The direction of Lomer-Cottrell dislocations is indicated by the direction of the hatching lines (according to Seeger et al. [66]).

From Fig. 23 it is possible to determine qualitatively the relative values of shear stresses in different slip systems. The smaller the distance between the point representing the axis of elongation and the stereographic triangle representing a definite slip system, the greater the shear stress acting in this slip system.

The generation of an appreciable number of Lomer-Cottrell dislocations in a given crystallographic direction requires a sufficient number of moving dislocations of the type necessary for their formation. This can occur if the acting stresses in the respective slip systems are higher than the critical stresses (taking into account the effect of hardening).

The transition from region I to region II occurs when the stresses become sufficiently large to create a sufficient number of dislocations in the secondary system, which leads to the formation of Lomer-Cottrell dislocations in the primary system. The stress necessary depends on the temperature, since the dislocations moving in the secondary system must intersect the forest of dislocations of the primary system. Since the slope of the hardening curve in region I is small, the dependence of τ_2 on T leads to a considerable dependence of ε_2 on T, which has been observed in experiments (see Fig. 23). Figure 23 also makes it possible to understand the dependence of γ_2 on orientation (represented, for example, in Figs. 11 and 12). The value of γ_2 is maximum when the direction of the axis of the sample coincides with or is close to [110], i.e., to the angle of the stereographic triangle. In the region coordinate with the (011) plane there is a tendency to form Lomer-Cottrell dislocations, represented in Fig. 22 by solid vertical lines. For this to occur the dislocations must be present in only one of the secondary systems—the conjugate system. If the elongation axis is close to [110] there is a tendency to form Lomer-Cottrell dislocations, represented by horizontal solid or dotted lines. Regardless of the orientation of the elongation axis with respect to the [110] direction in the triangle, the tendency toward the formation of Lomer-Cottrell dislocations of this type decreases, since the acting shear stress will decrease in at least one of the two slip systems which contain the necessary dislocation pairs. The end of the transition from stage I to stage II is reached when sufficient numbers of Lomer-Cottrell dislocations are formed in all three possible slip directions.

These processes are virtually insensitive to the degree of perfection of an undeformed crystal, the concentration of impurities in the crystal, or the diameter of the sample, which has been experimentally confirmed.

The investigations mentioned above do not exhaust the recent investigations of the effect of orientation, temperature, and deformation rate on the parameters of the hardening curves of monocrystals in terms of the dislocation theory. Koehler [87], Friedel [85, 88], Feltham and Meakin [89, 90, 91], and other investigators consider similar problems and in some cases solve them differently from the authors we have cited. However, the investigations we have discussed reflect the state of the problem sufficiently well, showing that many complex phenomena in the plastic deformation of crystals can be explained by dislocation theory. On the other hand, the complexity of the phenomena and the indeterminacy of the initial conditions (initial state of the crystal) make it possible to explain the same facts in somewhat different ways. Therefore at the present time the concept of dislocation can be applied essentially only as a working hypothesis. Only experiments can decide which of them are correct and which are erroneous. In this review it is unnecessary to enumerate all the working hypotheses—it would make this survey prohibitively long. The purpose of citing the investigations mentioned here was to emphasize certain achievements in the application of the dislocation theory to explain complex laws of the kinetics of the plastic deformation of monocrystals.

CONCLUSION

Some authors investigating the dislocation theory of the plasticity of crystals and the theory of viscous flow of amorphous bodies have concluded that the deformation mechanism is essentially different in crystals and amorphous bodies. Kochendörfer [92], for example, came to this conclusion in his survey of the basic laws of flow in amorphous and crystalline bodies. According to Kochendörfer the dislocation mechanism of plasticity is peculiar to crystals, in which, contrary to amorphous bodies, there is a long-range order in the distribution of atoms. The dislocation mechanism is characterized by a strong dependence of the yield point on the stress and a weak dependence on the deformation rate and temperature. Viscous flow is characterized by a strong dependence of the resistance to deformation on temperature and deformation rate.

As we have mentioned before, other authors emphasize, on the contrary, the fact that various phenomena and thermodynamic relationships determining the laws of plastic deformation are common to solids with very different structures. For example, Betcherman [93] and Korovskii [94] relate the temperature and time dependence of the yield point to the general laws of activation processes independently of the structure of the solid. Korovskii, on the basis of a purely thermodynamic approach, emphasizes that the activation theory is applicable to bodies of any structure.

On the other hand, although Korovskii emphasizes the important role of activation processes he evaluates incorrectly the role of the dislocation theory, saying that the dislocation theory does not take into account the collective character of regroupings, which determine the process of plastic deformation in crystals. In reality the dislocation theory of plasticity explains the mechanism of collective regroupings (mutual displacements of large regions of the crystal) as the result of the displacement of dislocations.

It seems to us that neither the dislocation nor activation theory should be singled out. These theories do not contradict each other and neither of them should be ignored.

In considering the plastic deformation of monocrystals and the relatively low dependence of some characteristics on temperature and deformation rate (e.g., yield point) one should not completely ignore the possible effect of activation processes. Under certain conditions (e.g., for a sufficiently large range of deformation rates) this effect must exist. From this viewpoint the assertion that one can consider critical stresses of the yield point type without taking the time factor into account is meaningless.

In considering the mechanism of plastic flow of amorphous and high-molecular substances it is necessary to take into account the role of collective regrouping, and consequently the possibility of the occurrence of athermic processes (not dependent on temperature or deformation rate) depending greatly on the stress. From this point of view not only the basic assumptions of the activation theory but also the dislocation theory of the plasticity of crystals becomes useful.

In other words, to develop the theory of plasticity of crystals it is impossible not to take into account the activation theory, and in the development of the theory of the plasticity of amorphous bodies it is necessary to take into account certain aspects of the dislocation theory.

Experimental and theoretical investigations of the temperature and time dependence of the strength and plasticity of solids with very different structures will help eliminate the contradictions in the results from the various investigations mentioned here and will help create a general physical theory of the strength and plasticity of solids.

LITERATURE CITED

1. P. P. Kobeko, Amorphous Substances [in Russian] (Izd. AN SSSR, 1952).
2. A. P. Aleksandrov, "Low-temperature resistance of high-molecular compounds," Reports 1 and 2, Conference on High-Molecular Compounds [in Russian] (Izd. AN SSSR, 1945) pp. 45-49.
3. Yu. S. Lazurkin and R. L. Fogel'son, "On the nature of large deformations of high-molecular substances in the glassy state," Zhur. tekh. fiz. 21, No. 3, 267-286 (1951).
4. Yu. S. Lazurkin, Mechanical Properties of Polymers in the Glassy State [in Russian] Doctoral dissertation (Moscow, 1954).
5. G. I. Gurevich, "On the laws of deformation of solids and liquids," Zhur. tekh. fiz. 17, No. 12, 1491-1503 (1947).
6. G. I. Gurevich, "The relationship between elastic and residual deformation in the general case of a uniformly stressed state," Trudy Geofiz. Inst. Akad. Nauk SSSR No. 21 (148), 49-90 (1953).
7. S. N. Zhurkov and B. N. Narzullaev, "Time dependence of the strength of solids," Zhur. tekh. fiz. 23, No. 10, 1677-1689 (1953).
8. S. N. Zhurkov and É. E. Tomashevskii, "The strength of solids. II. The effect of stress on longevity," Zhur. tekh. fiz. 25, No. 1, 66-73 (1955).
9. S. N. Zhurkov and T. P. Sanfirova, "Temperature and time dependence of the strength of pure metals," Doklady Akad. Nauk SSSR 101, 237-240 (1955).
10. S. N. Zhurkov, "Problems of the strength of solids," Vest. Akad. Nauk SSSR, No. 11, 78-82 (1957).

11. S. N. Zhurkov and T. P. Sanfirova, "Relationship between strength and creep in metals and alloys," Zhur. tekh. fiz. 28, No. 8, 1719-1726(1958).

12. I. E. Dorn, "Some fundamental experiments on high-temperature creep," J. Mech. and Phys. Solids 3, No. 2, 85-116 (1955).

13. L. M. Shestopalov, "Calculation of relaxation stresses in metals," Zhur. tekh. fiz. 26, No. 5, 1021-1033 (1956).

14. W. Kauzmann, "Flow of solid metals from the standpoint of the chemical-rate theory," Trans. Amer. Inst. Mining Eng. 143, 57-83 (1941).

15. C. M. Zener and J. H. Hollomon, "Problems of inelastic deformation of metals," Usp. Fiz. Nauk 31, No. 1, 16-37(1947). Transl. from J. Appl. Phys.

16. Collection: Elasticity and Inelasticity in Metals [Russian translation](IL, 1954).

17. Ya. I. Frenkel', "Temperature dependence of plasticity and creep in crystals," Zhur. éksp. teor. fiz. 9, No. 10, 1238-1244 (1939).

18. E. Schmid and V. Boas, Plasticity of Crystals, Particularly Metal Crystals [Russian translation] (GONTI, 1938).

19. A. V. Stepanov and E. A. Mil'kamanovich, "Optical limit of elasticity in rock salt crystals in the case of slip along the plane of the cube," Zhur. éksp. teor. fiz. 18, No. 9, 769-772 (1948).

20. A. V. Stepanov and V. P. Bobrikov, "Effect of temperature on the optical limit of plasticity along the (111) [011] system in rock salt crystals," Zhur. tekh. fiz. 26, No. 4, 795-799 (1956).

21. Theile, "Temperaturabhängigkeit der Plastizität und Zugfestigkeit von Steinsalzkristallen," Z. Phys. 75, 763-776 (1932).

22. D. I. McAdam and R. W. Mebs, "The technical cohesive strength and other mechanical properties of metals at low temperatures," Trans. Amer. Soc. Test. Materials 43, 661-703 (1943).

23. E. L. Vogel and R. M. Brick, "Deformation of ferrite single crystals," J. Metals 5, Sect. II, No. 5, 700-706 (1953).

24. T. Jokobory, "Delayed yield and strain rate and temperature dependence of yield point in iron," J. Appl. Phys. 25, No. 5, 593-594 (1954).

25. H. W. Paxton and I. J. Bear, "Further observations on yield in single crystals of iron," J. Metals 7, Sec. II. No. 9, 989-994 (1955).

26. R. Maddin and N. K. Chen, "Plasticity of molybdenum single crystals at high temperature," J. Metals 6, Sect. II, No. 2, 280-284 (1954).

27. I. H. Bechtold, "Strain rate effects in tungsten," J. Metals 8, Sec, II. No. 2, 142-146 (1956).

28. C. S. Barrett, "Effects of temperature on the deformation of beta-brass," J. Metals 6, Sect. II, No. 9, 1003-1008 (1954).

29. R. E. Jamison and F. A. Sherill, "The critical shear stress in α-brass as a function of zinc concentration and temperature," Acta metallurg. 4, No. 2, 197-200 (1956).

30. N. N. Davidenkov and T. N. Chuchman, "Survey of the modern theory of low-temperature fracture," Investigations of Heat-Resistant Alloys [in Russian] (Izd. AN SSSR, 1957) Vol. 2, pp. 9-33.
31. V. R. Regel', "Elongation curves of TlBr— TlI monocrystals," Tr. Inst. Kristall. No. 11, 152-157 (1955).
32. V. R. Regel' and A. B. Zemtsov, "Effect of crystallographic orientation of TlBr— TlI monocrystals on the flow limit during elongation tests," Tr. Inst. Kristall. No. 11, 158-164 (1955).
33. V. R. Regel' and G. E. Tomilovskii, "Dependence of the flow limit of TlBr— TlI monocrystals on the temperature and deformation rate," Tr. Inst. Kristall. No. 12, 158-171 (1956).
34. V. R. Regel', "Effect of temperature and deformation rate on the flow limit of monocrystals," Investigations of Heat-Resistant Alloys [in Russian] (Izd. AN SSSR, 1957) Vol. 2, pp. 275-280.
35. T. H. Blewitt, R. R. Coltman, and I. K. Redman, "Work-hardening in copper crystals," Report of a Conference of Defects in Crystalline Solids (Phys. Soc. London, 1955) pp. 369-382.
36. T. H. Blewitt, R. R. Coltman, and I. K. Redman, "Low-temperature deformation of copper single crystals," J. Appl. Phys. 28, No. 6, 651-660 (1957).
37. Z. S. Basinski and A. Sleesmyk, "On the ductility of iron at 4.2°K," Acta metallurg. 5, No. 3, 176-179 (1957).
38. T. S. Noggle and J. S. Koehler, "Electron microscopy of aluminum crystals deformed at various temperatures," J. Appl. Phys. 28, No. 1, 53-62 (1957).
39. J. S. Erikson and J. R. Low, "The yield-stress temperature relation for iron at low temperature," Acta metallurg. 5, No. 7, 405-406 (1957).
40. R. I. Garber, I. A. Gindin, V. S. Kogan, and B. G. Lazarev, "The plastic properties of beryllium monocrystals," Fiz. Metal. i Metalloved. 1, No. 1, 529-537 (1955).
41. V. R. Regel' and V. G. Govorkov, "The effect of temperature and deformation rate on the critical shear stress," Kristallografiya 3, No. 1 (1958).
42. V. D. Kuznetsov, Solid State Physics [in Russian] (1941) Vol. 2.
43. M. A. Bol'shanina, "Hardening and relaxation on the basis of plastic deformation," Izvest. Akad. Nauk SSSR, Ser. Fiz. 14, No. 2, 223-231 (1950).
44. M. A. Bol'shanina, "Effect of temperature on the mechanical properties of materials," Tr. SFTI 32, 170-174 (1953).
45. L. I. Vasil'ev, "Quasi-relaxational theory of rest," Tr. SFTI 26, 107-115 (1948).
46. L. I. Vasil'ev, "Annealing of Metals, I," Zhur. tekh. fiz. 20, No. 5, 619-628 (1950).

47. L. I. Vasil'ev, "Annealing of Metals. II," Zhur. tekh. fiz. 22, No. 11, 1827-1833 (1952).

48. L. I. Vasil'ev, "Annealing of Metals. III," Zhur. tekh. fiz. 23, No. 8, 1394-1399 (1953).

49. L. I. Vasil'ev, A. S. Bylina, and M. P. Zagrebennikova, "Effect of the change of the deformation rate on plastic elongation," Doklady Akad. Nauk SSSR 90, No. 5, 767-770 (1953).

50. V. R. Regel' and G. A. Dubov, "The kinetics of the annealing of TlBr–TlI monocrystals," Kristallografiya 2, No. 6, 756-759 (1957).

51. I. Weertman, "Creep of polycrystalline aluminum as determined from strain rate tests," J. Mech. and Phys. Solids 4, No. 4, 230-234 (1956).

52. P. Feltham, "On the mechanism of high-temperature creep in metals with special reference to polycrystalline lead," Proc. Phys. Soc. B, 69, 1173-1188 (1956).

53. P. Feltham, "On the activation energy of high-temperature creep in metals," Phil. Mag. 2, No. 17, 584-588 (1957).

54. I. Weertman, "Theory of steady-state creep based on dislocation climb," J. Appl. Phys. 26, No. 10, 1213-1217 (1955).

55. I. Weertman, "Compressional creep in tin single crystals," J. Appl. Phys. 28, No. 2, 196-197 (1957).

56. I. Weertman, "Steady-state creep through dislocation climb," J. Appl. Phys. 28, No. 3, 362-364 (1957).

57. V. R. Regel', V. G. Govorkov, and G. F. Dobrzhanskii, "Effect of temperature and deformation rate on the parameters of elongation curves of AgCl monocrystals," Zhur. Optich. Prom. No. 6, 38-42 (1958).

58. J. Gilman, "Plastic anisotropy of zinc monocrystals," J. Metals 8, No. 10, 1326-1336 (1956).

59. G. Masing, "Raffelsieper. Mechanische Erholung von Aluminium-Einkristallen," Z. Metallkunde 41, No. 3, 65-70 (1950).

60. P. Haasen and G. Leibfried, "On the meaning of the orientation dependence of the strain-hardening for aluminum single crystals," Z. Phys. 131, No. 4, 538-543 (1952).

61. K. Lücke and W. Staubwasser, Über die Temperaturabhängigkeit der Verfestigung von Al-Einkristallen," Naturwissenschaften 41, No. 3, 60(1954)

62. B. Jaoul and P. Lacombe, "Forme des courbes de traction de monocristaux d'aluminium raffiné," Compt. rend. 240, No. 25, 2411-2413 (1955).

63. E. N. Andrade and C. Henderson, "The mechanical behavior of single crystals of certain face-centered cubic metals," Phil. Trans. Roy. Soc. London A 244, No. 880, 177-203 (1951).

64. E. N. Andrade and D. A. Aboav, "The mechanical behavior of single crystals of metals in particular copper," Proc. Roy. Soc. A, 240, No. 1222, 304-320 (1957).

65. J. Diehl, "Zugverformung von Kupfer-Einkristallen. I. Verfestigungskurven und Oberflächenerscheinungen," Z. Metallkunde 47, No. 5, 331-343 (1956).

66. A. Seeger, J. Diehl, S. Mader, and H. Rebstock, "Work-hardening and work-softening of face-centered cubic metal crystals," Phil. Mag. 2, Sec. 8, No. 15, 323-350 (1957).

67. A. H. Cottrell and R. I. Stokes, "Effects of temperature on the plastic properties of aluminum crystals," Proc. Roy. Soc. A, 233, No. 1192, 17-34 (1955).

68. M. A. Adams and A. H. Cottrell, "Effect of temperature on the flow stress of work-hardened copper crystals," Phil. Mag. 46, No. 382, 1187-1193 (1955).

69. G. Schoeck, "Dislocation theory of plasticity of metals," Advances appl. Mech. 4, 229-279 (1956).

70. V. L. Indenbom, "Mobility of dislocations in the Frenkel'-Kontorova model," Kristallografiya 3, No. 2, 197-205 (1958).

71. A. Seeger, "Theorie der Kristallplastizität. I. Grundzüge der Theorie; II. Die Grundstruktur der dichtest gepackten Metalle und ihr Einfluss auf die plastische Verformung; III. Die Temperatur und Geschwindigkeitabhängigkeit der Kristallplastizität," Z. Naturforsch. 9a, 758-775, 856-881 (1954).

72. E. H. Edwards, J. Washburn, and E. R. Parker, "Some observations on the work-hardening of metals," Trans. Amer. Inst. Mining and Metallurg. Eng. 197, 1525-1529 (1953).

73. A. H. Cottrell and B. A. Bilby, "Dislocation theory of yielding and strain aging of iron," Proc. Phys. Soc. A, 62, No. 349, 49-62 (1949).

74. Fisher, Trans. Amer. Soc. Metals 47, 451 (1955).

75. G. L. Pearson, W. T. Read, and W. L. Feldman, "Deformation and fracture of small silicon crystals," Acta metallurg. 5, No. 4, 181-191 (1957).

76. H. D. Dietze, "Die Temperaturabhängigkeit der Versetzungsstruktur," Z. Phys. 132, No. 1, 107-110 (1952).

77. A. Seeger, "Das Fliessen der Metalle bei hohen Temperaturen," Z. Metallkunde 45, No. 9, 521-527 (1954).

78. A. Seeger, "The temperature dependence of the critical shear stress and of work hardening of metal crystals," Phil. Mag. 45, No. 366, 771-773 (1954).

79. A. Seeger, "The generation of lattice defects by moving dislocations and its application to the temperature of FCC crystals," Phil. Mag. 46, No. 382, 1194-1217 (1955).

80. A. Seeger, "Stacking faults in close-packed lattices," Report of a Conference on Defects in Crystalline Solids (Phys. Soc. London, 1955) pp. 328-339.

81. G. I. Taylor, "The mechanism of plastic deformation of crystals. Pt. I. The theoretical," Proc. Roy. Soc. A, 145, 368-387 (1934).

82. N. F. Mott, "A theory of work-hardening and metal crystals," Phil. Mag. 43 , No. 346, 1151-1178 (1952).

83. G. Leibfried and P. Haasen, "Zum Mechanismus der plastischen Verformung," Z. Phys. 137, 67-88 (1954).

84. G. Schoeck and A. Seeger, "Activation energy problems associated with extended dislocations," Report of the Conference on Defects in Crystalline Solids (Phys. Soc. London, 1955) pp. 340-346.

85. I. Friedel, "On the linear work-hardening rate of face-centered cubic single crystals, "Phil. Mag. 46, No. 382, 1169-1186 (1955).

86. H. W. Paxton and A. H. Cottrell, Work-hardening in stretched and twisted aluminum crystals," Acta metallurg. 2, No. 1, 3-8 (1954).

87. J. S. Koehler, "Theory of initial stress-strain curves in face-centered metals," Acta metallurg. 1, No. 3, 377 (1953).

88. I. Friedel, "The mechanism of work-hardening and slip-band formation," Proc. Roy. Soc. A, 242, No. 1229, 147-159 (1957).

89. P. Feltham and J. D. Meakin, "On the mechanism of work-hardening in face-centered cubic metals, whith special reference to polycrystalline copper. I," Phil. Mag. 2, No. 13, 105-112 (1957).

90. P. Feltham and J. D. Meakin, "Work-hardening in FCC metal crystals," Acta metallurg. 5, No. 10, 555-564 (1957).

91. P. Feltham and J. D. Meakin, "On the mechanism of work-hardening in face-centered cubic metals with special reference to polycrystalline copper. II," Phil. Mag. 2, No. 22, 1237-1245 (1957).

92. A. Kochendörfer, "Die Grundgesetze des Fliessens der amorphen und der kristallinen Stoffe," Z. angew. Phys. 5, No. 2, 69-80 (1953).

93. I. I. Betcherman, "Rate processes in physical metallurgy, Progr. Metal. Phys. 2, 53-89 (1950).

94. Sh. Ya. Korovskii, "Basis of the activation theory of plastic deformation and fracture of solids," Tr. Rizhskogo Inzh. Aviaychilishcha No. 34, 3-29 (1957).

ADDITIONAL LITERATURE

1. J. J. Gilvarry, "Temperature-dependent equations of state of solids," J. Appl. Phys. 28, No. 11, 1253-1261 (1957).

2. Z. S. Basinski, "Activation energy for creep of aluminum at subatmospheric temperatures," Acta metallurg. 5, No. 11, 684-686 (1957).

3. I. W. Suiter and W. A. Wood, "Deformation of magnesium at various rates and temperatures," J. Inst. Met. 91, No. 4, 181-188 (1952).

4. J. H. Bechtold, "Tensile properties of annealed tantalum at low temperatures," Acta metallurg. 3, No. 3, 249-254 (1955).

5. G. Jastram, Plastische Verhalten von Kupferdrähten bei verschiedenen Dehnungsgeschwindigkeiten. Arbeitstag der Festkörperphysik (Dresden, 1954) Vol. 2, pp. 82-93.

6. F. X. Eder and H. I. Wisotzki, "Der Einfluss der Verformungsgeschwindigkeit auf Lage und Ausbildung der Streckgrenze bei Stahl," Z. Metallkunde 48, No. 10, 561-564 (1957).

7. N. Loizou and R. B. Sims, "The yield stress in pure lead in compression," J. Mech. and Phys. Solids 1, No. 4, 234-243 (1953).

8. D. S. Clark, "The behavior of metals under dynamic loading," Trans. Amer. Soc. Metals 46, 34-62 (1954).

9. N. P. Aelen, B. E. Hopkins, and I. E. McLennom, "The tensile properties of single crystals of high-purity iron at temperatures from 100-253°C," Proc. Roy. Soc. A, 234, No. 1197, 221 (1956).

10. E. T. Wessel, "Some exploratory observations of the tensile properties of metals at very low temperatures," Trans. Amer. Soc. Metals 49, 149-169 (1957).

11. F. F. Vitman and V. A. Stepanov, "Effect of deformation rate on the cold brittleness of steel," Zhur. tekh. fiz. 11, No. 12, 1070-1084 (1939).

12. L. D. Sokolov, "Effect of deformation rate on the resistance of metals to plastic deformation," Zhur. tekh. fiz. 16, No. 4, 437-442 (1946).

13. L. D. Sokolov, "Static and impact compression of brasses containing different amounts of zinc," Zhur. tekh. fiz. 16, No. 11, 1277-1282 (1946).

14. F. F. Vitman, "Effect of deformation rate on the cold brittleness of steel. 3," Zhur. tekh. fiz. 17, No. 1, 77-86 (1947).

15. N. N. Davidenkov and A. V. Noskin, "A high-speed drop hammer for bending and elongation," Zab. Lab. 13, No. 6 (1947).

16. N. N. Davidenkov, "Summary of 30 years of mechanical testing in the USSR," Zav. Lab. 13, No. 11, 1330-1346 (1947).

17. Ya. Potak, "Method of determining the susceptibility of quenched steels to spontaneous destruction under the effect of static stress," Zav. Lab. 13, No. 1, 77-84 (1947).

18. L. D. Sokolov, "Effect of deformation rate and temperature on the resistance of metals and amorphous bodies to plastic deformation," Zhur. tekh. fiz. 17, No. 5, 543-548 (1947).

19. F. F. Vitman, N. N. Davidenkov, N. A. Zlatin, and B. S. Ioffe, "Use of conical impressions in the body of the effect of deformation rate on the resistance of metals to deformation," Zav. Lab. 14, No. 5, 579-594 (1948).

20. L. D. Sokolov, "Effect of grain size on the relationship between the rate of deformation and stress during plastic deformation," Zhur. tekh. fiz. 18, No. 1, 89-92 (1948).

21. L. D. Sokolov, "Effect of degree of deformation on the relationship between stress and rare of deformation," Zhur. tekh. fiz. 18, No. 1, 93-97 (1948).

22. L. D. Sokolov, "The rate coefficient in different representations of the stressed state," Zhur. tekh. fiz. 18, No. 5, 687-696 (1948).

23. L. D. Sokolov, "Resistance to plastic deformation at high deformation rates," Zhur. tekh. fiz. 18, No. 5, 697-700 (1948).

24. F. F. Vitman, N. A. Zlatin, and B. S. Ioffe, "Resistance of metals to deformation at the rate of 10^{-6}-10^2 m/sec. I," Zhur. tekh. fiz. 19, No. 3, 300-314 (1949).

25. F. F. Vitman and N. A. Zlatin, "Resistance of metals to deformation at rates of 10^{-6}-10^2 m/sec. II," Zhur. tekh. fiz. 19, No. 3, 315-326 (1949).

26. F. F. Vitman and N. A. Zlatin, "Resistance of metals to deformation at rates of 10^{-6}-10^2 m/sec. III," Zhur. tekh. fiz. 20, No. 10, 1267-1272 (1950).

27. F. F. Vitman, N. A. Zlatin, and L. M. Shestopalov, "Relationship between activation energy and resistance to deformation of metals," in Collection Dedicated to A. F. Ioffe on his 70th Birthday [in Russian] (Izd. AN SSSR, 1950) pp. 331-340.

28. N. N. Davidenkov, F. F. Vitman, and N. A. Zlatin, "Effect of aging on the rate and temperature dependence of hardness," in Collection Dedicated to A. F. Ioffe on his 70th Birthday [in Russian] (Izd. AN SSSR, 1950) pp. 307-317.

29. L. D. Sokolov, "Theory of the effect of deformation rate on the resistance of metals to plastic deformation," Zhur. tekh. fiz. 20, No. 4, 447-457 (1950).

30. L. D. Sokolov, "A systematic investigation of deformation rate and temperature dependence of the resistance of monophase metals to deformation," Doklady Akad. Nauk SSSR 70, No. 5, 839 (1950).

31. N. N. Davidenkov and S. Belyaev, "Hardening steel by precipitation of carbides," Zhur. tekh. fiz. 22, No. 1, 40-43 (1952).

32. N. A. Zlatin and N. Ya. Nikolenko, "Effect of deformation rate, temperature, and heat treatment on the hardness of medium carbon steel," Zhur. tekh. fiz. 22, No. 10, 1565-1571 (1952).

33. A. Noskin, V. Delle, A. Moiseev, and B. Plisov, "Testing metals with a universal high-speed drop hammer," Zav. Lab. 18, No. 8, 989-994(1952).

34. G. I. Pogodin-Alekseev and S. V. Zhuravlev, "Effect of deformation rate on plasticity characteristics in the case of elongation," Sb. Nauch. Rabot Mosk. Inzh-Fiz Inst. No. 8, 108-119 (1954).

35. A. M. Rozenberg and A. N. Eremin, "Effect of the deformation rate on stress in the process of cutting plastic metals," Izvest. Tomsk. Politekhn. Inst. 75, 26-46 (1954).

36. N. F. Kunin and A. T. Sukhin, "Plastic compression of metals at different temperatures," Tr. Chelyabin. Inst. Mekh. i Elek. Sel'sk. Khoz. No. 5, 134-142 (1955).

37. L. I. Vasil'ev et al., "Effect of the rate and degree of plastic elongation on relaxation and subsequent deformability of metals," Fiz. Metal. i Metalloved. 2, No. 1, 142-145 (1956).
38. V. Voloshenko-Klimovitskii, "Method of measuring forces and small deformations during elongation by impact," Zav. Lab. 22, No. 9, 1090-1094 (1956).
39. Ya. B. Gurevich, "Character of the flow limit in the case of elongation," Fiz. Metal. i Metalloved. 2, No. 1, 137-141 (1956).
40. N. N. Davidenkov, "Summary of 40 years of mechanical testing in the USSR," Zav. Lab. No. 10, 1244-1265 (1957).
41. Z. A. Ridness and A. P. Chekmarev, "Effect of temperature and rate and degree of deformation on the resistance of carbon steel to plastic deformation," Izvest. Akad. Nauk SSSR, OTN No. 12, 22-30 (1957).
42. V. M. Savitskii, Effect of Temperature on the Mechanical Properties of Metals and Alloys [in Russian] (Izd. AN SSSR, 1957).
43. L. M. Shestopalov, Effect of the Rate and Period of Application of Stress on the Value and Distribution of Stresses and Plastic Deformations [in Russian] (Izd. AN SSSR, 1957).
44. B. N. Alexander, Z. Metallkunde, No. 5, 344-352 (1961).
45. V. G. Govorkov and V. R. Regel', "Effect of temperature and deformation rate on the parameter of the compression curves of germanium monocrystals," Kristallografiya 3, No. 5, 1324-1330 (1961).
46. V. G. Govorkov, "Effect of temperature on the shape of compression curves of silicon monocrystals," Kristallografiya 6, No. 5, 789-791(1961).
47. H. G. Van Bueren, Imperfection in Crystals (North Holland Publishing Company, 1960).

A. A. URUSOVSKAYA

PLASTIC DEFORMATION NOT INDUCING ASTERISM IN LAUE SPOTS

X-ray investigations of deformed crystals began shortly after Laue's discovery of x-ray diffraction by crystal lattices (1912). Rinne [1] observed that spots on the Laue diagram of deformed mica were elongated radially, i.e., the Laue spots showed asterims. The first systematic investigation of plastic deformation by the use of x-ray was made in the early 1920s by A. F. Ioffe, M. V. Kirpicheva, and M. A. Levitskaya [2], and from then on the x-ray method of investigation became the most frequently used method of studying the plastic deformation of crystals. The shape of the spots is an indication of the changes occurring in crystal lattices as a result of deformation and, consequently, is to some degree an indication of the process of deformation itself.

Ioffe, Kirpicheva, and Levitskaya investigated rock salt crystals deformed by slipping. On the basis of the fact that Laue spots are drawn out into tails (asterism), the authors concluded that asterism always accompanies deformation by slip. However, at the present time cases are known where slip is not accompanied by asterism. To determine the mechanism of the plastic deformation of crystals it is very important to study deformations of the lattice which do not induce elongation of Laue spots into tails.

CAUSES OF ASTERISM

First of all let us consider the causes of asterism. At the present time there are several hypotheses of the relationship between asterism and the process of plastic deformation. These hypotheses can be divided into three main groups:

1) Asterism is induced by local bending of the slip planes.

2) Asterism is induced by fragmentation of the lattice accompanying slip (formation of subblocks, etc.), and also by polygonization.

3) Asterism is induced by the reorientation of regions (incoherent twins, faults, deformation bands, etc.).

The hypotheses of the first group were developed by Yamaguchi [3], who considered that the slip planes become deformed when the process of slip is showed down by some cause. On the basis of their experimental results, Burgers and Lebbink [4] also concluded that local bends occur along the slip planes. These authors investigated slip in Al crystals subjected to shearing.

60

A special apparatus was constructed to induce this type of deformation; the shear stress was applied along the slip plane and along the direction of slip. Before and after deformation the samples were subjected to x-ray analysis. The x-ray photographs were taken from a rotating crystal (rotational x-ray photographs) and from a stationary crystal (Laue diagrams). On rotational x-ray photographs the spots did not change their location or shape after deformation. This result led to the conclusion that for practical purposes pure slip had occurred (translation). However, the Laue diagram of these samples showed asterism. Analysis of the Laue diagrams led to the conclusion that slip, even in the case of deformation by shearing (when bending of the sample is prevented), is accompanied by local bending of the slip planes. The axis of the bend is in the slip plane and is perpendicular to the direction of slip. The hardening occurring in these experiments was explained by the authors as caused by local bending. In the same investigation Burgers and Lebbink advanced the hypothesis of the relationship between the asterims of Laue spots and the number of acting slip planes. The greater the number of acting slip planes the less bending occurring along each of these planes and the less the asterism of Laue spots. The possibility is not excluded that the Laue method as well as the rotation method is not sensitive to the distortion of the lattice due to slip.

Manteuffel [5], Komar [6], Wood [7], Cahn [8], and others are proponents of the second hypothesis. These investigators have observed a discrete splitting of spots. On the basis of the fact that they observe discrete asterims during deformation at high temperatures they assume that such asterism is due to polygonization, i.e., the bends of the crystal planes become fractures.

Other authors [9] interpret asterism as the result of rotational microblocks due to the fragmentation of the material along the slip plane (the roller scheme).

The third theory, explaining asterism as the formation of regions with a rotated lattice, has been the most widely accepted in recent years. N. A. Brilliantov and I. V. Obreimov [10] were the first to point out this relationship. These authors observed that interlayers with the lattice rotated with respect to the initial orientation are formed during the compression of rock salt. The angle of rotation of the lattice was small, of the order of a few degrees, and depended on the degree of deformation. The interlayers were approximately parallel to the $\{110\}$ plane. The Laue spots of the deformed samples showed typical asterism. On the basis of their results Brilliantov and Obreimov assumed that there is in fact a translational displacement of crystal layers. The "slip" bands observed are the boundaries of rotated regions. These regions, called incoherent twins by the authors, are responsible for the asterism of the Laue spots.

Barrett and Levenson [11] studied the deformation bands on mono- and polycrystalline samples of different metals and came to the conclusion that the formation of regions with a rotated lattice during deformation is related to asterism. The compression of monocrystals, and also the rolling, stretching, and elongation of metal polycrystals (iron, brass, tin, etc.), leads to the formation of bands which, for small degrees of deformation, are located along the planes with small indices. For greater degrees of deformation the lattice in these bands becomes disoriented and the bands themselves change shape, becoming twisted. Metallographic and x-ray analyses gave the same series of orientations, and this was the basis for the assumption of the relationship between asterism and the presence of nonoriented regions. Similar conclusions were drawn by Calnan [12].

In this respect the investigation by Honeycombe [13] is very significant. Honeycombe subjected samples of Al monocrystals to stretching. The Laue diagram of the deformed samples always showed asterism, even when the degree of deformation was very low— about 1%. Each spot was composed of several maxima connected by a weak, diffused background. Comparison of metallographic and x-ray analyses showed that the diffused background corresponded to the fault bands, and the maxima to the disoriented region between fault bands and also to bands with secondary slips. The deformation in Al, and consequently the asterism, is not eliminated by annealing.

M. V. Klassen-Neklyudova and A. A. Urusovskaya [14] have also shown the existence of a relationship between asterism and the formation of reoriented regions. These authors studied fault bands in TlBr— TlI crystals. The fault bands (deformed regions of the crystal) represent an assembly of wedge-shaped regions oriented in different directions. If the x-ray beam used for the Laue diagram envelops the whole fault interlayer or a considerable part of it, then asterism occurs. If the beam passes into only one wedge shaped region, where there are no visible deformations of the lattice, then there is no asterism even if there are traces of slip.

SLIP WITHOUT ASTERISM

Incoherent twins, deformation bands, faults, etc., are macroscopic forms of plastic deformation occurring when a heterogeneous distribution of stresses accompanied deformation. It follows then that if the sample is deformed only by slip and is prevented from bending, asterism does not occur. Such experiments were made, in which samples were deformed in three ways: by applying shear stress along the slip plane parallel to the direction of slip (deformation by shear), by applying concentrated force to the crystal, and by careful elongation.

We shall briefly describe some of these experiments.

A. Kochendörfer [15], investigating plastic deformation of naphthalene monocrystals by three different methods of deformation (elongation, compression, shearing) observed that slip is not accompanied by asterism of the Laue spots in deformation of shearing. Slip occurring during compression and elongation does induce asterism of the Laue spots. On this basis Kochendörfer concluded that asterism is not connected specifically with slip. He found that slip occurring in deformation by shearing is accompanied by hardening and concluded, from this observation, that hardening due to slip and asterism are not necessarily concomitant.

Further experiments on deformation of Al crystals by shearing were made by Röm and Kochendörfer [16]. The samples were longer than those used by Burgers and Lebbink and a crystal layer 50 mm wide was subjected to deformation by shearing (in the work of Burgers and Lebbink the interlayer subjected to deformation was only 0.5 mm wide). The total degree of deformation was 50%. When deformation was 20-25% no traces of slip were visible on the surface of the sample. Laue diagrams of these samples showed no asterism (the diagrams are not given in the article). The shape of the deformation curve indicated the occurrence of work hardening. For greater degrees of deformation slip traces appeared on the surface of the sample and the Laue spots acquired tails, i.e., ordinary asterism was observed. If the greatly deformed samples were subjected to annealing (350° for one hour) there was complete relaxation, the asterism disappearing.

Aside from this investigation there are others in which slip was observed without any accompanying asterism of Laue spots. The investigation by Honeycombe [13] is devoted to a special study of the causes of elongation of Laue spots as the result of deformation. Honeycombe elongated cadmium and aluminum crystals. The cadmium crystals were elongated very carefully on a glass plate at room temperature by using needle type pincers. In this way he managed to elongate samples as much as 300%. Laue diagrams were made before and after elongation. No asterism was observed for samples deformed by 100-200%. However, careful observation of the inner parts of the Laue diagram showed a fine structure which disappeared after the samples were annealed. There was no recrystallization as the result of annealing. Typical asterism on the Laue diagrams of deformed cadmium monocrystals was observed only when they were deformed by bending or in cases there fault bands were formed during elongation. In the latter case asterism was observed when the Laue diagrams of the fault bands were made. After relaxation these samples showed polygonization and recrystallization.

M. V. Klassen-Neklyudova and A. A. Urusovskaya [17] also observed deformation without asterims in the case of TlBr— TlI crystals. Concentrated stress on these crystals lead to slip spreading through the crystal along the $\{110\}$ planes in $<100>$ directions. This slip leads to the appearance of "through impact and pressure figures" on the surface of the crystal.

The study of "through figures" in crystals of CsBr made it possible to determine the point at which slip not accompanied by asterism becomes slip accompanied by asterims. Figure 1 is a photograph (a CsBr crystal taken in
polarized light) of a surface parallel to the direction of the concentrated stress.
This surface is parallel to the (110) plane. The slip planes are perpendicular
to the plane of this surface. If one introduces an additional gypsum plate
when using polarized light then the light bands (traces of slip) are alternately

Fig. 1. Photograph in polarized light of a CsBr
sample with a "through pressure figure." The light
lines on the dark background are slip traces; the
pressure was applied along the [001] direction perpendicular to the upper surface of the sample; the
through figure appeared on the lower surface; the
plane of the surface (plane of the figure) is parallel
to the (110) plane (perpendicular to the slip planes
and parallel to the direction of slip).

red and green. On the basis of these results one can assume that compressed
and elongated regions occur in the slipped layers of the crystal. A Laue diagram of the slip region is shown in Fig. 2. There is no elongation of the spots
into tails, although the spots do not remain completely regular. They preserve
the size and general shape of spots from undeformed samples but acquire a
complex structure, becoming striated.

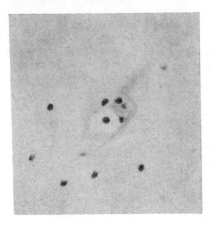

Fig. 2. Laue diagram of CsBr sample
shown in Fig. 1. The photograph was
taken at an angle of 4 degrees from
the [110] direction.

The Laue diagram of the same samples after annealing shows the same striations, which indicates that the fine structure in the spots is caused by disoriented regions in the slip traces.

The results described lead us to conclude that if the slip occurring in the crystal is not accompanied by deformation of the slip planes when such slip does not lead to asterism of Laue spots. The deformations of the lattice appearing along such slip traces are insufficient to change the diffraction pattern. Asterism appears only if the slip process is accompanied by macroscopic deformation resulting from deformation of lattice planes, with the formation of rotated regions (polygonization blocks, faults, etc.). The shape of the spots changes only when the reflecting capacity of the crystal in macroscopic volumes— of the order of a few microns and higher— is changed.

LITERATURE CITED

1. T. Rinne, Ber. Sächs. Acad. Wiss., Math. phys. Kl. 17, 223 (1915).
2. A. F. Ioffe, M. V. Kirpicheva, and M. A. Levitskaya, "Deformation and strength of crystals," Zhur. Russ. fiz. khim. obsh. 4, 489 (1924).
3. K. Yamaguchi, "The internal strain of uniformly distorted aluminum crystals," Sci. pap. Inst. Phys. Chem. Res. (Tokyo) 11, 151 (1929).
4. W. Burgers and F. Lebbink, "Lattice distortion in sheared aluminum single crystals," Rec. trav. chim. 64, 321 (1945).
5. I. F. Manteuffel, "Die Entwicklung des Asterismus in Steinsalzkristallen," Z. Phys. 70, 109 (1931).

6. A. P. Komar, "The structure of plastically deformed crystals according to Laue-patterns. I," Phys. Z. Sowjetunion 9, 413 (1936).

7. W. Wood, "Crystalline structure and deformation of metals," Proc. Phys. Soc. A, 52, 110 (1940).

8. R. Cahn, "Recrystallization of single crystals after plastic bending," J. Inst. Metals 70, 121 (1949-1950).

9. A. F. Ioffe, Lectures on Molecular Physics [in Russian] (Leningrad, 1923) p. 259.

10. N. A. Brilliantov and I. V. Obreimov, "Plastic deformation of rock salt. III," Zhur. éksp. teor. fiz. 5, No. 3-4, 330-339 (1935); Plastic deformation. IV Zhur. éksp. teor. fiz. 7, No. 8, 978, 986 (1937).

11. C. S. Barrett and L. Levenson, "Structure of iron after drawing, swaging and elongation in tension," Trans. AIMME 135, 327-343 (1939).

12. E. Calnan, "Laue asterism and deformation bands," Acta crystallogr. 5 557-564 (1952).

13. R. Honeycombe, "Inhomogeneities in the plastic deformation of metals. I," J. Inst. Metals 80, 45 (1951-1952).

14. M. V. Klassen-Neklyudova and A. A. Urusovskaya, "The structure of fault bands in crystals of thallium halides," Kristallografiya 1, No. 5, 564-571 (1956).

15. A. Kochendörfer, Plastische Eigenschaften von Kristallen und metallischen Werkstoffen (Berlin, 1941) s. 12.

16. F. Röm and A. Kochendörfer, "Neue Ergebnisse über die Verfestigung bei der plastischen Verformung von Kristallen," Z. Metallkunde 41, 265 (1950).

17. M. V. Klassen-Neklyudova and A. A. Urusovskaya, "Through impact and pressure figures of cubic halide crystals," Trudy Inst. Kristallografiya Akad. Nauk SSSR 11, 140-145 (1955).

A. A. URUSOVSKAYA

FORMATION OF REGIONS WITH A REORIENTED LATTICE AS A RESULT OF DEFORMATION OF MONO- AND POLYCRYSTALS

According to the classical theory, plastic deformation of crystals can occur either by slipping or twinning.

According to an idealized concept, slip consists in the simultaneous displacement of parallel layers of the crystal along definite crystallographic planes in a definite direction—the direction of slip. One part of the crystal is displaced with respect to another part by a certain distance which is a multiple of a lattice period. Thus the internal structure is not distorted as the result of slipping but steps occur on the surface of the deformed crystal.

Mechanical twinning consists in the reorientation of large areas of the crystal; it follows one of the two laws of twinning: either reflection in the plane of twinning, or rotation by a definite angle (180 or 360°) around a definite crystallographic axis. The indices of the twinning plane or the twinning axis usually have a very small value and are strictly determined for each crystal. The lattices of the two components should coincide completely in their plane of separation.* In the case of reflection twins the contact plane between two individuals is the plane of symmetry.

In 1885, O. Lehman [1] observed in a number of mineral crystals the presence of regions in which the lattice was rotated at an angle with respect to the original lattice of the crystal. The angle of rotation of the lattice was not constant and depended on the degree of deformation. Many investigators have considered these regions as twins but O. Mügge [2] was the first to propose differentiating the irregularly oriented regions (with respect to the rest of the crystal) from the regularly oriented regions—twins, which he called Knick-bander. During the past five or six years a great number of investigations of such regions in metal crystals has been published. These publications

---

*If a transitional layer with a structure intermediate between the structure of both individuals does not occur between the components of the twin. Such transitional layers exist, for example, in the case of wurtzite twins.

indicate that the phenomenon of rotation of the lattice plays a very important role in the plastic deformation of monocrystals as well as polycrystals.

The purpose of this article is to give a short historical review of investigations of the various manifestations of deformation in which the crystallographic lattice is rotated under the effect of mechanical action.

REORIENTED REGIONS IN MINERALS

O. Mügge [2] presented the first detailed description of bands with the lattice rotated with respect to the rest of the crystal. He observed such bands in anhydrite, antimonite, kyanite, vivianite, gypsum, mica, graphite, molybdenum, barium chloride, sodium nitrate, barite, and calcite. Figure 1a shows a mica crystal deformed by compression in which can be seen three such bands, and Fig. 1b shows the position of the lattice in a similar interlayer in

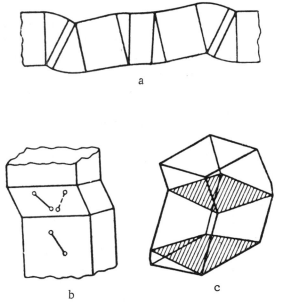

Fig. 1. Crystals in which bands with a rotated lattice occurred as the result of compression. Drawings by Mügge. a) Mica; b) kyanite; c) calcite.

kyanite. According to these schemes the lattice of the interlayer is not in a twinning position with respect to the lattice of the original crystal. The reorientation of the lattice in the calcite crystal resulting from compression along one edge of the crystal is shown in Fig. 1c. The rotation of the lattice occurs near the ends of the sample. Mügge showed that deformation accompanied by rotation of the lattice is characteristic of many crystals but he did not make any detailed investigation of the mechanism of the process. However, he made the assumption that the observed bands result either from complicated slip or twinning.

R. Brauns [3] observed a "twinned" striation along the noncoherent edges of the natural faces of rock salt. His careful investigation of the samples showed that the striations correspond to interlayers in which the lattice is rotated with respect to the neighboring part of the crystal. According to Brauns' results the angle of rotation of the lattice in these bands is 4° and the "twinning" plane is the plane with an index of (20 · 20 · 1) which is located in the immediate vicinity of the (110) planes of the rhombododecahedron. Brauns notes that he could not reproduce such regions by mechanical means.

The appearance of large regions with a rotated lattice in rock salt crystals as the result of mechanical action was first observed by N. A. Brilliantov and I. V. Obreimov [4]. These investigators subjected rock salt crystals cut along the habit plane to compression along the short side. The relative dimensions of the samples were 1 : 10 : 10, with a height of 2-4 mm. As the result of compression the edges of the samples became corrugated with little facets. The angles between neighboring facets, measured by both optical and x-ray methods, varied from a few minutes to a few degrees. The separation

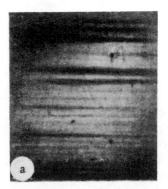 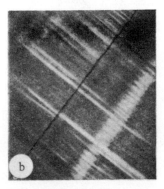

Fig. 2. Noncoherent twins in rock salt. a) Photograph of the side of the deformed sample in reflected light; b) photograph of the same sample in polarized light; the light lines visible in polarized light correspond to the boundaries between facets (Brilliantov and Obreimov).

planes of neighboring regions were not constant. Their angle of deviation from the (110) plane was ± 1.5°. A photograph with light reflected from the edge surface of a compressed rock salt crystal is shown in Fig. 2a. A photograph of the same side of the sample in polarized light is shown in Fig. 2b. The light lines, reflecting the internal stresses, correspond to the boundaries of the facets. These lines are boundaries between rotated regions of the crystal. In Fig. 3 is shown Brilliantov and Obreimov's representation of the structure of the sample having an interlayer with a rotated lattice. According to this scheme the boundary of the interlayer does not correspond to any crystallographic plane, i.e., a plane with small indices. The deformations of the structure are concentrated along this boundary but the crystal is not deformed within the interlayer. Within the interlayer the lattice is oriented symmetrically with respect to the rest of the crystal so that the boundary of the inter-

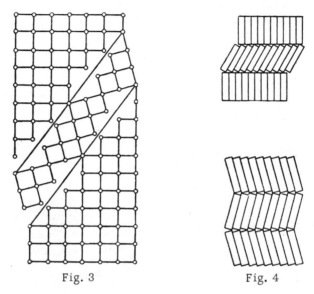

Fig. 3   Fig. 4

Fig. 3. Diagram of the structure of the sample with a noncoherent twin according to Brilliantov and Obreimov. The noncoherent twin is formed by the rotation of the lattice.
Fig. 4. Diagram of the formation of noncoherent twins by slip (the second scheme by Brilliantov and Obreimov).

layer is a plane of symmetry of contiguous lattices. Thus Brilliantov and Obreimov consider the facets occurring from deformation of rock salt to be twins with a twinning plane which is not any constant plane. They call these "noncoherent twins." Brilliantov and Obreimov were the first to emphasize the important role of this type of deformation in the plastic flow of crystals. They indicate the necessity of a detailed investigation of this phenomenon to clarify the mechanism of plastic deformation in general. In their investigation they assumed that twinning along the "noncoherent" faces is the basic process leading to plastic deformation of crystals. Twinning along the coherent plane they consider to be the limit case of noncoherent twinning; they completely negate the possibility of translational slip and consider that the existence of such slip must be demonstrated.

Brilliantov and Obreimov also state that the rotation of the lattice can be considered as translational slip accompanied by a rotation of the slip planes (Fig. 4) but if this is so then slip within the rotated region would not show up, since during translational slip no deformation of the lattice should occur.

They negate the possibility of purely translational slip and therefore doubt such a representation of the formation of "noncoherent twins."

The angle of rotation of the lattice during the formation of noncoherent twins can vary widely, depending on the conditions of the experiment. Brilliantov and Obreimov deformed rock salt at room temperature and under this condition the angle of rotation of the lattice did not exceed four degrees. V. I. Startsev [5], compressing rock salt crystals at temperatures up to 400°C, obtained angles of rotation up to six degrees. The structure of the sample was transformed into blocks rotated with respect to each other.   By optical and x-ray methods he found a relationship between the number of blocks and the total angle of rotation of the lattice with respect to the original crystal as a function of the applied stress. The number of blocks increases with increasing stress. The total angle of rotation of the crystal lattice also increases.

Qualitatively, the variation of the rotation angle as a function of the applied stress is similar to the deformation curve (hardening curve, i.e., curve of the variation of stress as a function of the degree of deformation). Annealing did not change the structure of the sample (the disorientation did not disappear).

To clarify the phenomenon of twinning along the noncoherent faces, M. V. Klassen-Neklyudova [6] deformed rock salt crystals by compression at even higher temperatures (500-600°C) and higher stress. Under these conditions she observed "twins" with very large angles of rotation (up to 38°). The twins had the shape of plates or grains with irregular boundaries. In some cases the crystals were transformed into monocrystals with an orientation close to that of the twin, according to the spinel law. Klassen-Neklyudova interpret this reorientation as a process analogous to mechanical twinning. Further investigation showed, however (Klassen-Neklyudova and Urusovskaya [48]), that the reorientation of the lattice in rock salt samples deformed at high temperatures results from a very rapid recrystallization.

The formation of rotated regions in rock salt has been observed not only as the result of compression but also as the result of elongation. A. V.

Stepanov [7] subjected rock salt crystals to elongation in the direction of slip,
[110], at high temperatures (50-300°C) and observed regions of platelike shapes
limited by planes close to the planes of the rhombododecahedron. The plates
extended through the whole thickness of the sample and were somewhat sloped
with respect to the original crystal. From the outside this type of deformation
looked more like twinning than slipping. This phenomenon was investigated
and described in more detail by Stepanov and A. V. Donskoi [8]. Deformed
samples in reflected light show dark and light stripes and the plates extend to
the contiguous faces of the cube (Fig. 5). Figure 6 shows a Laue diagram
from the boundary between two individuals (two plates). One can easily see

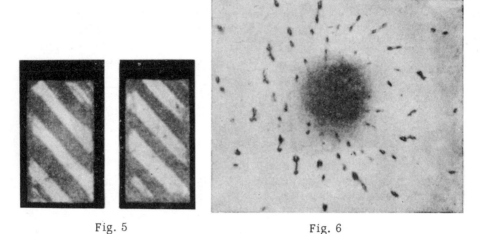

Fig. 5   Fig. 6

Fig. 5. Photographs (in reflected light) of two adjacent faces of a cube in a
rock salt sample deformed by layering (A. V. Stepanov [8]).

Fig. 6. Laue diagram of the boundary between two plates (A. V. Stepanov).

the splitting of the spots into two reflections which are connected by weak lines.
The shape of the Laue spots indicates that the lattice within the plate is rotated at some angle with respect to the original crystal. This angle of rotation is 12° or more.

The two neighboring plates do not form a twin. This was confirmed by
the following observation. Indices of the planes interconnecting the plates
were determined on the basis of the consideration that the faces of the plates
are the planes of the cube. It turned out that this plane has one index [e.g.,

(111)] with respect to one of the plates and another one [e.g., (355)] with respect to the other. On the basis of this observation the authors conclude that this phenomenon represents a special type of deformation which, according to their assumptions, has the same significance as slip or twinning. This formation of asymmetrically rotated regions (plates) was called "layering" by the authors.

Let us note two interesting facts observed by Stepanov and Donskoi in their investigation of deformation resulting in layering. After the formation of plates (at stresses of about 780 g/ mm^2) the structure of these plates changes with increasing stress. Bands—striations—occur in these plates and the striations become more and more clearly defined with increasing stress. Stepanov and Donskoi assumed that these striations represented traces of slip. The traces of slip lie parallel to each other in each plate and at a given angle to the boundary of the plates. The authors conclude that these traces of slip represent secondary phenomena, i.e., that slip becomes possible after the lattice has rotated into a position favorable for slip.

Their study of layering gave some indication that the lattice in the plates ceases to be cubic and becomes monoclinic with an angle of 30' to 10° between the x and z directions. The decrease of the degree of symmetry of the rock salt crystals to monoclinic as a result of plastic deformation was observed earlier by N. J. Seljakov [9], who subjected rock salt crystals to 8-10% compression and then to bending in the direction perpendicular to the [100] direction. The angle observed was one degree at a maximum.

KINK FORMATION IN METAL CRYSTALS RESULTING FROM DEFORMATION BY COMPRESSION

The investigations previously described were relative to minerals. Deformation accompanied by rotation of the lattice has been investigated in the greatest detail in rock salt, although such investigations are few and incomplete. A much greater number of investigations has been made of metal monocrystals. The article by E. Orowan [10], which appeared in 1942, concerns the compression of a cadmium bar in the direction parallel to the slip plane; it has served as the starting point of the study of the deformation of interlayers with rotated lattices in metals. During compression the sample underwent longitudinal steplike bending (Fig. 7) and a particular interlayer was formed which Orowan identified as the same Knick-bander observed by Mügge in minerals. Orowan called this interlayer a kink band, which means a band with a break (i.e., a break in the lattice). In Russian this interlayer is more conveniently a "fault band."

In geology the mutual displacement of crystal blocks is called a fault. In crystals also the upper part is displaced with respect to the lower and between

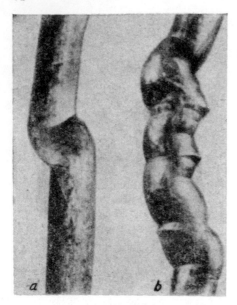

Fig. 7. Kinks in cadmium (a) as the result of slight compression and (b) as the result of considerable compression (Orowan [10]).

the displaced parts there is a transition zone which is the "interlayer," or kink band. In what follows we shall call the kneelike bends resulting from deformation "kinks" (see Fig. 7).

Orowan gives a scheme of the formation of kink bands (Fig. 8). Under the effect of shear stress slip occurs in the crystal just above the boundary of the kink band, KK, and this part rotates from its original position but is not a mirror image of the lower part of the crystal. This is indicated by the arrows in Fig. 8, which show the lattice orientation of the original crystal (lower part of the figure) and of the kink band (upper part of the figure).

Orowan considers the noncoherent twins of Brilliantov and Obreimov to be kinks, and the formation of a rotated region to be a new type of deformation of crystals, as did Stepanov. However, he considers the reorientation of the lattice to be the result of slip.

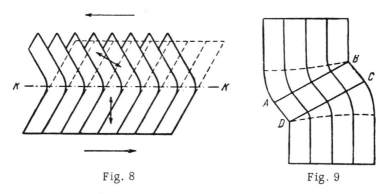

Fig. 8   Fig. 9

Fig. 8. Diagram of the formation of kinks according to Orowan.

Fig. 9. Diagram of the structure of a kink band (Hess and Barrett [11]).

J. Hess and C. Barrett [11] have assumed, however, that the data given by Orowan are insufficient to ascertain that kink formation is a particular case of plastic deformation. Hess and Barrett investigated zinc, which has a structure similar to that of cadmium, to make sure whether kink bands are formed there and whether this type of deformation can be reduced to phenomena already known, i.e., ordinary slip, slip with bending, or twinning. These authors subjected bars of monocrystalline zinc to compression; the basal plane (slip plane) made an angle of three degrees with the axis of the bar. Under this condition kink bands were formed very easily. The boundary of the kink band was approximately perpendicular to the basal plane and was not displaced during the process of deformation. These authors ascertained that kink bands are formed in zinc only if the sample is fixed rigidly on both sides so as to prevent rotation. Otherwise a continuous longitudinal bend occurs, accompanied by twin interlayers. The kink bands were formed in different places: at the edges, at a distance from the edge of the sample of one-third or one-fourth of the length. Kinks appeared at the edges only when the surface of the sample had no defects whatsoever. If the surface had some local defects (inclusions, scratches, etc.) the kinks were always localized at these defects. Within the kink bands the lattice of the crystal was rotated and the angle of rotation increased with an increasing degree of compression; in some cases the angle of rotation reached $88°$. Zinc samples with kinks were cracked and found to have a smooth bend of the basal plane within the kink; the axis of rotation was in the basal plane and was perpendicular to the direction of slip $[\bar{2}110]$ close to the direction of compression at the beginning of the formation of the kink. With increasing degrees of deformation the axis of rotation deviated from this position and tended to become perpendicular to the axis of the sample. The boundaries of the kink bands were not crystallographic boundaries.

Figure 9 represents the structure of the kink band formed at an early stage of deformation. According to this scheme the kink band at the beginning consists of a rotated but not bent part of the crystal—the parallelogram ABCD—to which two symmetrical regions with bent slip planes are adjacent. During the process of deformation the part of the crystal located within the ABCD region also becomes bent. Since the formation of kinks is accompanied by bending of slip planes, the authors consider kink formation to be a particular case of ordinary slip with bending, and not a specific type of deformation.

Hess and Barrett give a qualitative explanation of the appearance of kink bands by using the dislocation theory. They assume that in the beginning plastic deformation occurs within the kink band and does not spread to the edges of the sample, i.e., the deformation begins with nucleation of pairs of dislocations of opposite signs within the crystal. The positive and negative dislocations accumulate within two parallel planes almost perpendicular to the slip plane (boundaries AB and CD in Fig. 9). The nucleation of dislocation

pairs is explained on the basis of an analysis of the data of the elasticity theory for the compression of a long isotropic bar. According to this theory, a
loss of the elastic stability of the bar occurs at a given compression stress and
the bending of the bar follows a cosinusoid. In the bends the consinusoids of
stress are maxima, and when these stresses become equal to the critical shear
stresses pairs of dislocations of opposite signs appear in all slip planes, assem
bling in flat series (Fig. 10a). At each of these planes the basal plane bends.
The basal plane cuts in two the angle between the slip planes located on each
side of it. With increasing degrees of deformation the planes containing the
dislocations of opposite signs tend to diverge and thus increase the width S of

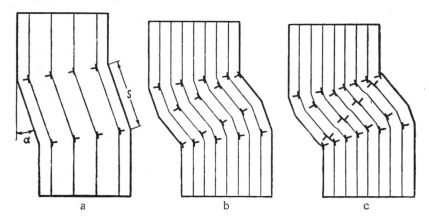

a   b   c

Fig. 10. Diagram of the formation of kinks by dislocations (Hess and Barrett).

the kink. The motion of the dislocation walls continues until they meet the
defects on the surface. After this the dislocation walls form the final boundaries of the kink. On further compression the angle α of the rotation of the
lattice within the kink band increases and secondary kink boundaries can occur within the kink band; these secondary kink boundaries are new planes with
dislocations, at which the basal planes become deformed (Fig. 10b, c). Finally, the slip planes in the whole kink band become bent.

In 1954 Gilman [12] published an investigation of kinks in zinc deformed
by compression. Some of his data are in agreement with those of Hess and
Barrett: kinks are formed in samples in which the principal axis makes a small
angle with the basal plane; the boundary of the kink is approximately perpendicular to the slip planes and cuts in two the angle between the planes located
at each side of this boundary. However, some of Gilman's data contradict the
assumptions of Hess and Barrett. Gilman found that to obtain kinks it is not
necessary to fix both sides of the sample rigidly during compression. He ob-

tained kinks with equal success in samples with edges rigidly held and in samples whose edges could move freely in cone-shaped holders. Furthermore,
Gilman did not succeed in producing kinks in samples which were compressed
strictly along the basal plane. On the basis of this fact he concludes that preliminary elastic longitudinal bending is not essential for the formation of kinks
He showed that compression followed by elongation can induce displacement
of the kink boundaries. He succeeded in eliminating kinks obtained during
compression by subsequent elongation of the sample. Gilman does not agree
with Hess and Barrett that the bending of slip planes accompanied kink formation. He considers that if a zinc sample is compressed along the basal plane
(when the motion of the dislocation along the slip plane is impeded) the basal
planes break instead of bending and the breaks occur in places where the shear
stress is minimum.

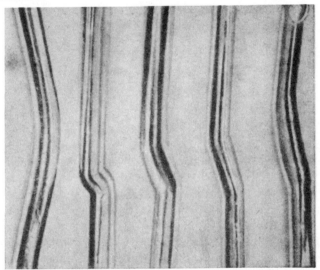

Fig. 11. Effect of the angle between the slip plane and the
axis of the sample on the width of the kink band in compressed zinc (Gilman [12]).

Gilman investigated the effect of the crystallographic orientation of the
sample on the formation of kinks, i.e., the effect of the χ angle between the
basal plane and the axis of the sample. His experiments showed that kink
bands do not occur during compression of samples in which the χ angle is
close to zero. As the angle is increased the number of cases where kinks occur increases; they appear most often in samples in which χ angle is 20-24°.
The width of the kink also increases with an increasing χ angle. This fact is
illustrated in Fig. 11, in which are represented several samples of deformed

crystals of zinc with basal planes differently oriented with respect to the axes
of these samples. In the first sample the basal plane is parallel to its axis.
This sample bends smoothly during compression. The other samples in the
figure are arranged in order of increasing χ angle, from 2.5 to 24°. It can be
seen that in the first samples the kink bands cover a smaller part of the crystal
than in the following crystals, and that in the last sample it is not even sharply defined.

Gilman made deformation curves during the process of compression
(curves of the variation of the degree of compression as a function of the applied stress). At the moment the kink is formed the stress drops sharply and a
sharp minimum occurs on the curve. The shape of the deformation curve corresponding to kink formation is similar to that corresponding to twinning.

We shall not describe other investigations concerning kink formation resulting from deformation by compression, but note only that kinks resulting
from compression have been observed in titanium [44] and other crystals.

We shall now describe the results of investigations of kinks in metals
which occur as the result of elongation.

KINK FORMATION IN METAL CRYSTALS RESULTING FROM DEFORMATION BY ELONGATION

The investigation by A. V. Stepanov and A. V. Donskoi showed that elongation of rock salt and silver chloride crystals may result in the formation of
regions with rotated lattices. Other authors have obtained kinks as the result
of elongation of crystals of iron [13], aluminum [29], tin [14], zinc [15, 16],
bismuth [45], magnesium [17], titanium [18], and copper [46].

A. Holden and F. Kunz [13] investigated kinks in iron crystals elongated
in the [111] direction of slip. The stress drops sharply the moment a kink
band occurs. The boundaries of the kink band coincide approximately with
the (111) plane. X-ray analysis showed that the lattice in the kink band is
rotated with respect to the rest of the crystal and the angle of rotation increases with the degree of deformation. Wiggly slip traces corresponding to
the (112) planes appear in the band and adjacent regions. Holden and Kunz
assumed that when cubic crystals are subjected to elongation kinks can form
when one slip system is acting and when slipping is impeded at some point in
the sample. The deformation curves obtained by these authors were similar
to those obtained by Gilman for kink formation in zinc as the result of compression.

Kink bands resulting from elongation are much clearer in zinc monocrystals than in iron. Figure 12 shows a typical zinc sample with kink bands
formed as the result of elongation. This figure is taken from the publication
by Regel' and Govorkov [19].

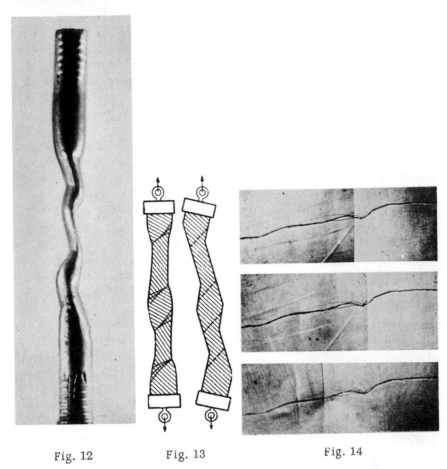

Fig. 12   Fig. 13   Fig. 14

Fig. 12. Zinc sample in which kink bands were formed as the result of elongation (Regel' and Govorkov [19]). $\chi = 62°$; temperature of the experiment = 200°C.

Fig. 13. The Washburn and Parker diagram explaining the occurrence of kink bands as the result of elongation.

Fig. 14. Displacement of the kink boundary as the result of stress (according to Washburn and Parker).

J. Washburn and E. Parker [15] also investigated kink formation resulting from elongation of zinc. Monocrystalline zinc bars were elongated under constant stress at high temperature. The angle between the geometric axis and the slip plane of the samples varied but was near 45°. Washburn and Parker assumed that the kink bands were caused by the irregular distribution of slip along the sample, i.e., the elongation due to slip is greater in one place than another. Possibly the clamps also had an effect. Bending moments are created due to the irregular distribution of stress at the boundaries of regions where slip has occurred. In these regions the lattice is bent and thus a kink band is formed. This mechanism of the formation of kinks is illustrated in Fig. 13, which was taken from the publication by Washburn and Parker. These authors also observed displacement of the boundary of the kink band in a zinc plate, which is illustrated in Fig. 14.

The rotation of the lattice as the result of the elongation of zinc crystals was also investigated by Gilman and Read [16], who made a special investigation of the effect of surface defects on plastic deformation. Scratches were made on the surfaces of samples with very different orientations (λ = 15-75°); the samples had a triangular cross section. As the result of elongation localization of plastic deformation and formation of a kink band occurred next to the scratch. The authors conclude that kinks occur because this hardened region blocks the slip. Bending moments are created at the boundary between this region and the rest of the crystal and the slip planes become bent. In this investigation two new "bend-planes" were observed. If the scratch made on the surface of the sample is transverse, then a wedge-shaped piece of the sample is partially separated from the sample during elongation. On further elongation the wedge-shaped piece bends away from the rest of the crystal(Fig. 15). If a longitudinal scratch is made on the surface of the sample then elongation leads to the following phenomenon: the slip planes rotate during elongation, tending to become parallel to the axis of elongation. These samples have two regions: in one region the slip planes rotate to one side, and to the opposite side in the other. As the result the triangular cross section acquires a more complex shape (Fig. 16). On the basis of these experiments Gilman and Read

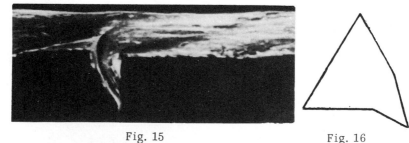

Fig. 15   Fig. 16

Fig. 15. Sample of zinc monocrystal in which a wedge-shaped piece separated at a transverse scratch as the result of elongation (Gilman and Read[16]).

Fig. 16. Shape of the cross section of a zinc crystal with a longitudinal scratch after elongation (Gilman and Read).

assume that the necessary condition for the occurrence of kinks in zinc crystals is the presence of local surface defects. The kinks are found just at those points.

DEFORMATION BANDS

Plastic deformation of a crystal induces not only regions with an asymmetrically rotated lattice of the kink-band type, plates, and noncoherent twins but also, as has been noted in the literature, deformation bands which have a different appearance from those described. Pfeil [20] was the first to describe such deformation bands, but Barrett has made a much more detailed investigation of them. Barrett subjected mono- and polycrystalline samples of α-iron to compression. As the result of compression, bands which did not look like slip traces appeared on the surfaces of the crystals (in separate grains in polycrystalline samples). At first the bands were not very clear, but with increasing degrees of compression they stood out much more clearly.   Figure 17 shows two photographs taken from Barrett's investigation; they show

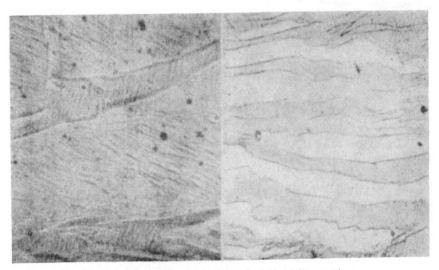

Fig. 17. Deformation bands in iron (Barrett).

surfaces of an iron sample in which deformation bands were formed. The photograph at the left represents a low degree of deformation and that on the right a greater degree of deformation. As can be seen, these bands have irregular boundaries and systems of thinner bands are found within each band. Barrett assumes that these thinner bands are also deformation bands.

X-ray investigation of deformation bands showed that the lattice in these bands is rotated at a certain angle which depends on the degree of compression. A model of the crystal structure containing a deformation band was drawn on the basis of the x-ray data. On this model (Fig. 18) the orientation of the lattice in different parts of the crystal is shown by differently oriented cubes. The lattice within the band is oriented in such a way that the [111] direction is parallel to the axis of compression, which coincides with the normal to the figure. The orientation of the lattice in the adjacent part of the sample is such that the axis of compression coincides with the [100] direction. Between these two regions the orientation of the lattice is intermediate.

Barrett explains the occurrence of deformation textures in metals on the basis of deformation bands. He observed deformation bands not only in iron but also in α and β-brass, tungsten, copper, silver, magnesium, and other alloys.

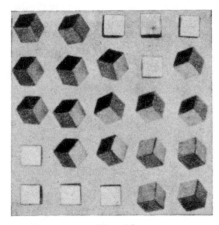 

Fig. 18   Fig. 19

Fig. 18. Model of the structure of the deformation band and the adjacent region (Barrett).

Fig. 19. The initial stage of formation of a deformation band in an aluminum crystal (Friedel).

J. Herenguel and P. Lelong [22] observed a similar phenomenon in rolled mono- and polycrystalline AlMg alloys. Bands became visible in the grains of polycrystals after etching. These authors explain the bands as reorientations of the lattice resulting from plastic deformation. Etching of rolled AlMg monocrystals revealed groups of dark and light bands in four different directions. The etching figures differ in shape and number in neighboring bands. It was demonstrated that the lattice in dark bands is in a twinned position with respect to the lattice in the light bands, with a common axis, [111], and there is a gradual change from one etching figure to another. This transition indicates a gradual rotation of the crystal lattice from one band to another. Apparently the bands observed by Herenguel and Lelong are similar to the deformation bands observed by Barrett.

Figure 19 is a photograph of a deformation band in its initial stage of formation in an aluminum crystal [23]. This photograph allows one to conclude that deformation bands result from the impedance of slips at the boundary of some region. This region becomes the deformation band. The reorientation of the lattice in the deformation band with respect to the lattice in the mother crystal occurs because during the process of deformation the lattice in the mother crystal rotates as the result of slipping while such rotation does not occur in the deformation band.

ACCOMMODATION BANDS

There is still another phenomenon which occurs as the result of plastic deformation of crystals and involves rotation of the lattice. D. S. Jillson[24] observed bands adjacent to twin interlayers in deformed zinc crystals. Jillson called these "accommodation bands," considering that the formation of a twin induces the deformation of the adjacent region, i.e., forces the surrounding material to accommodate itself to the existence of the twin. Jillson observed the formation of accommodation bands at a linear boundary between two grains. However, he could not reproduce this phenomenon.

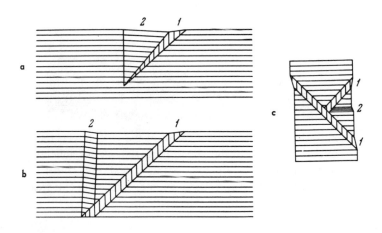

Fig. 20. Different means of accommodation to the existence of a twin according to Jillson. a) Accommodation to the presence of a wedge-shaped twin interlayer by bending of the basal planes; b) accommodation of the crystal to the existence of a twin interlayer both of whose limiting planes are twinning planes by the bending of the basal plane; c) accommodation at two twin interlayers by slipping along the basal plane; 1) twin interlayer; 2) accommodation region.

Accommodation occurs either by bending of slip planes or by slipping in the basal plane. In the latter case there is no rotation of the lattice. Up to now there is no experimental proof of the existence of such a process of accommodation. Figure 20 is a schematic representation of different mechanisms of the formation of accommodation bands according to Jillson. If the twin does not extend through the whole sample but is stopped by some defect or by a boundary between grains, then the region of the accommodated material is localized next to the twin.

A. Moore [25] has also found accommodation bands in zinc. These bands appeared next to the twin interlayer between regions in which the basal plane of the original crystal had changed its direction (Fig. 21a). Moore measured the angle of the change in direction on a profilogram; it was 47 ± 2.5'.

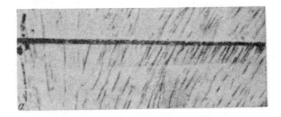

Fig. 21. Accommodation bands at the twin interlayer
in zinc. a) One of the accommodation planes adheres
to the twin interlayer; b) two accommodation planes
adhere to the twin interlayer on one of its sides (Moore).

Sometimes the twin interlayer was connected to two successive accommodation bands (Fig. 21b) with different angles with respect to the original crystal: the angle of one band, closer to the twin interlayer, was 21', and that of the other, further from the interlayer, was 14'.

In later investigations Moore [26] studied the accommodation bands in the cross section of a zinc crystal perpendicular to a trace of a twin interlayer on the basal plane. The surface of the cross section was mechanically ground

and polished and then electrolytically polished. The surface was examined
in reflected polarized light. A micrograph of a cross section is shown in
Fig. 22. One can see the cross section of the wedged twin (dark triangle) and
the accommodation band next to it. The micrograph coincides with the
scheme proposed by Jillson (Fig. 20a).

V. I. Startsev [50, 51] and Pratt and Pugh [27] have investigated accommodation bands in greater detail. Startsev observed a fine structure of the
accommodation band and the presence of a strongly deformed region at the
boundary between the accommodation band and the mother crystal, i.e., the
existence of a secondary transition region. By selective etching he investigated the distribution of dislocations in the accommodation band in bismuth
monocrystals. He also investigated the effect of annealing on the primary
and the secondary accommodation bands. This investigation showed that
there is a close relationship between the formation of accommodation bands
and deformation by slip. Pratt and Pugh [27] investigated habit planes of zinc
split in liquid air. As the result of splitting different traces of plastic deformation occur on the habit plane: slip lines, twin interlayers, and accommodation bands. Figure 23 is a micrograph of accommodation bands formed
next to the twin interlayer. The micrograph shows that aside from accommodation bands parallel to the twin, there are also bands perpendicular to and
at an angle to the twin.

It was shown that the width of the accommodation band parallel to the
twin traces depends on the width of the twin itself: if the twin interlayer nar-

   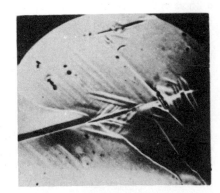

Fig. 22   Fig. 23

Fig. 22. Photograph in reflected polarized light of an accommodation region
at a wedged-in twin interlayer on the cross section perpendicular to the trace
of the band on the basal plane (Moore).

Fig. 23. Photograph of accommodation bands next to the twin interlayer in
zinc (Pratt and Pugh).

rows at the end then the accommodation band disappears at the end. A scheme similar to that proposed by Jillson is given to explain the appearance of bands parallel to the twin interlayer.

To explain the occurrence of accommodation bands at the end of the wedged-in twin interlayer, perpendicular to the interlayer, or at an angle of 30° to it, the following reasoning, based on the crystallography of the habit plane of zinc, is given: on the habit plane, which is the basal plane, the atoms are packed closest to each other. The directions of the closest packing are the $<1\bar{2}10>$ directions of slip. There are three directions of slip. In these directions there are traces of twin interlayers which have occurred in the $\{10\bar{1}2\}$ planes of the first order pyramid. It is well known that in kink formation the basal planes rotate around an axis in the basal plane which is perpendicular to the direction of slip. Since in zinc crystals the direction of slip coincides with the trace of the twin interlayer on the basal plane, the kink bands on the habit plane must be perpendicular to the $<1\bar{2}10>$ traces of twins. At the extremity of the wedged-in twin interlayer there are always stresses, which can be resolved by the formation of bands with a rotated lattice.

Accommodation bands parallel to twin interlayers are called primary, and those perpendicular to them are called secondary. Pratt and Pugh consider accommodation bands to be related to kinks.

BANDS OF SECONDARY SLIP

Deformation resulting in rotated regions has been investigated by many authors and the phenomenon and its causes have been interpreted in many different ways. Barrett [21, 28] and Jackson and Chalmers [14] consider that the asymmetric reorientation of the material resulting from plastic deformation occurs because slip follows different systems of crystallographic planes in contiguous regions of the crystal. Calnan [29], Holden and Kunz [13], and Gilman and Read [16] assume that bands with a rotated lattice are formed if slipping is possible only in one system of slip planes (for whatever reason) and only if this slip is impeded by some region, which becomes the kink band. Honeycombe's investigation [30] settled this difference of opinion. Using optical and x-ray methods, he made a detailed investigation of anomalous traces of plastic deformation in cadmium and aluminum subjected to elongation, and concluded that there are two types of plastic deformation which result in rotation of the lattice: kink bands induced by deformation of the acting slip planes, and bands in which slip occurs along a system of planes different from the system of planes along which other parts of the sample have slipped. These latter Honeycombe called bands of secondary slip. The position of these bands is different from that of kink bands. Kinks are almost perpendicular to the acting slip traces in the matrix while bands of secondary slip are approximately parallel to the main slip traces. Figure 24a is a mi-

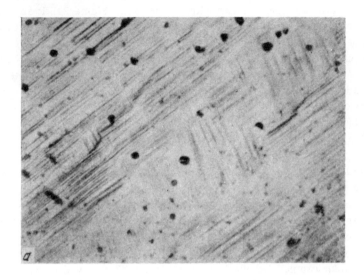

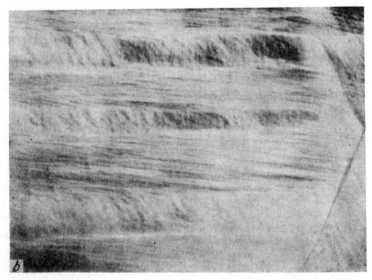

Fig. 24. Bands of secondary slip in elongated aluminum. a) Traces of secondary slip in the band are linear; b) traces of secondary slip are irregular (Honeycombe).

crograph taken from Honeycombe's investigation of a region of the aluminum sample where bands of secondary slip can be seen. Figure 24b is another micrograph from Honeycombe's work where the bands have a somewhat different appearance because the secondary slip traces are irregular.

Kink bands as well as bands of secondary slip induce asterism of Laue spots. All investigators agree that in kink formation the lattice rotates around an axis in the slip plane and perpendicular to the direction of slip. Laue diagrams of the kink band show splitting of spots corresponding to one rotation axis. Laue diagrams of bands of secondary slip show much more complex splitting. However, Calnan [29] succeeded in determining one axis of rotation of the lattice in a band of secondary slip. This axis of rotation was perpendicular to the acting slip planes. Slipping accompanied by the rotation of the slip planes around the normal to the slip planes was investigated in detail by Wilman [31] and was called rotational slip. Thus one can assume that the occurrence of bands of secondary slip is induced by slip along the secondary system of slip planes and that this slipping is accompanied by rotation of slip planes around the normal to these planes.*

Wilman assumes that the "deformation bands" observed by Barrett [21] in mono- and polycrystalline iron subjected to compression are one of the manifestations of rotational slip.

KINKS IN THALLIUM AND CESIUM HALIDE CRYSTALS

A number of investigations concerning deformation accompanied by formation of rotated regions have been made by this author under the direction of M. V. Klassen-Neklyudova. In these investigations [32, 33, 34, 35] kink bands in TlBr–TlI, CsI, and CsBr were investigated. The structure of kink bands and the experimental conditions which result in the formation of kink bands were investigated. From our experiments we obtained data on the mechanism of the formation of kinks and the relationship between the processes of kink formation and slip.

Crystal of thallium and cesium halides are very convenient for the study of kink bands because their transparency makes it possible to observe the structure of the kinks in transmitted polarized light. Furthermore, an impact with a needle induces characteristic impact figures which make it possible to determine the orientation of the lattice directly without the use of x-rays. Also, slip in TlBr–TlI, CsI, and CsBr crystals can be impeded considerably because in crystals with a structure of the CsCl Type the slip planes are parallel to the edges of the cube. When these crystals are oriented in such a way

---

* From the viewpoint of dislocation theory rotational slip occurs as the result of displacement of screw dislocations. Slip accompanied by a rotation of the lattice around an axis located in the slip plane and perpendicular to the direction of slip is due to the displacement of a series of edge dislocations.

that slip is impossible in one crystallographic direction then slip in other slip directions is automatically excluded.

Let us first examine the conditions under which kinks are formed. To elucidate this problem we subjected samples with different crystallographic orientations to compression and elongation. Usually compression of elongation of TlBr− TlI, CsI, and CsBr crystals induces slip along one or several slip systems, {110} <100>. However, if one creates conditions which impede slip, then kink bands are formed as the result of deformation. This can be done by using samples oriented in a specific way. In the case of compression

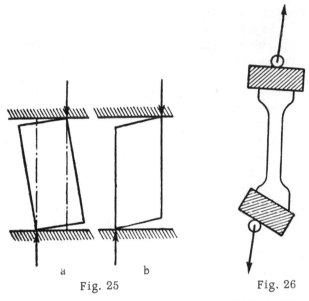

a   b
Fig. 25   Fig. 26

Fig. 25. Diagram of the experiment to obtain kinks in TlBr−TlI, CsI, and CsBr crystals to obtain kinks.
Fig. 26. Diagram of the experiment to obtain kinks in TlBr− TlI crystals as the result of elongation (simultaneous elongation and bending).

kinks occur most often in samples compressed in directions close to the [100] direction of slip. In the case of elongation kinks are formed in samples elongated in the [110] direction; (if the sample is elongated in the [100] direction then brittle fracture occurs, and if elongated in the [111] direction the result is complex slip).

Kinks may occur not only as the result of impeding slip by using different orientations of the crystal but also as the result of irregular distribution of stress. Experiments showed that irregular distribution of stress favors the occurrence of macrobending moments in separate areas of the sample. Such a distribution of stress occurs when the sample is compressed in such a way that the edges are at an angle to the face of the press (Fig. 25a) or the edges of the sample are not parallel (edges poorly made), (Fig. 25b). A diagram of the experimental setup to induce kinks by elongation is shown in Fig. 26. In this case elongation is accompanied by bending of the sample.

Aside from the effect of macrobending stresses on the formation of kinks, the effect of local stress concentrations was also investigated. These experiments were made with metal monocrystals [10, 15] and it was shown that kinks always occur in overstressed areas (at scratches, pits, etc.). Experiments with thallium and cesium halide crystals containing holes, pits, scratches, etc. showed [33] that the presence of stress concentrations does not induce formation of kinks with either compression or elongation. These results contradict the results of the experiments made with metals. The difference in the effect of local defects in metals on one hand and thallium and cesium halide on the other can be explained on the basis of the effects of plastic deformation on crystals of different substances (with very different types of bonds). Experiments showed [35, 36, 37] that stress concentrations induce deformation by slip in TlBr— TlI, CsI, and CsBr crystals while in metal crystals (zinc, bismuth, tin) it induces twin interlayers and bands adjacent to them in which the lattice is slightly rotated with respect to the original crystal (accommodation bands analogous to kinks). Thus the process of plastic deformation in areas of stress concentrations in crystals of thallium and cesium halide is different from the deformation process under analogous conditions in metals.

On the basis of these results we may assume that plastic deformation in TlBr— TlI, CsI, and CsBr monocrystals results in the formation of kinks if an irregular distribution of stresses induces a complex stressed state. Aside from this, it is necessary that the crystallographic axes be oriented in a definite way with respect to the deformation axis of the sample.

To clarify the mechanism of the reorientation of the crystal lattice in kink bands we investigated the structure of these bands in a great number of thallium and cesium halide samples. The structure of the kinks was examined directly under a microscope (in transmitted, polarized, or ordinary light) or by use of Laue diagrams. In all CsI and CsBr samples, and some of the TlBr— TlI samples, there are slip traces within the kink bands. Figure 27a is a photograph of part of such a kink band in a TlBr— TlI crystal, and Fig. 27b that in a CsBr crystal. The photographs show that slip traces passing through the kink band gradually change their direction. Slip traces in TlBr— TlI crystals change direction by steps at the boundaries of separate wedge-shaped regions, in which the kink band is broken, but remain linear and parallel to each other within the limits of each wedge-shaped region. One of the boundaries of each wedge-shaped region is perpendicular to the slip traces within

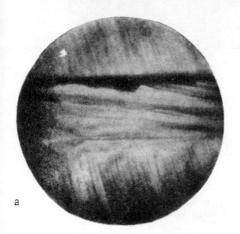

a   b

Fig. 27. Photographs of kink bands: a) In a TlBr—TlI crystal (in ordinary transmitted light); b) in a CsBr crystal (in polarized light).

this region. In a CsBr crystal the slip traces indicate that the lattice in the kink band is rotated with respect to the rest of the crystal. The rotation of the lattice was also confirmed by x-ray analysis. It was shown that the [110] direction is the axis of the rotation of the lattice in the formation of kinks. This direction is in the slip plane and is perpendicular to the direction of slip.

The boundaries of kink bands in different samples are crystallographically different and in some cases have very high indices. However, the kink bands have a tendency to orient themselves approximately perpendicular to the acting or possible planes and directions of slip. This tendency is particularly well-defined when the rate of compression is very high or when the degree of deformation is very high.

The presence of slip traces within kink bands leads us to assume that the reorientation of the lattice as the result of deformation is induced by slip accompanied by rotation of the slip planes. However, another assumption is possible, namely that slips existed either before the reorientation process or occurred after the formation of the kink. To elucidate the role of slip in the process of reorientation of a region of the lattice it was necessary to determine whether it is possible to separate the process of slip from the process of rotation of the lattice. We succeeded in obtaining rotation of the lattice in TlBr—TlI monocrystals without visible slip traces as the result of compression of samples of a definite crystallographic orientation, or strictly along the slip planes, or in a direction at a small angle from the [100] direction. Figure 28 represents photographs of two sections of the TlBr—TlI crystal compressed exactly in the [100] direction. The photographs were taken with transmitted ordinary light. The TlBr—TlI crystals have the property that the slip traces

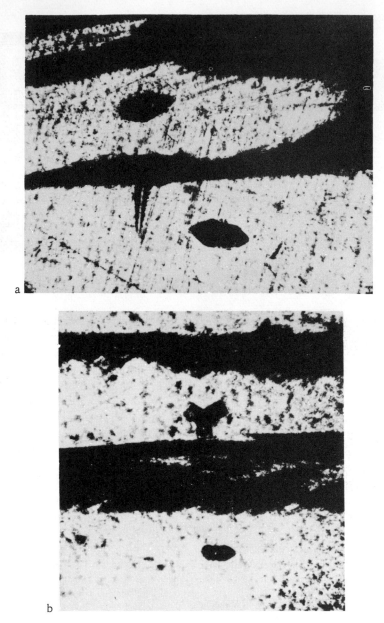

Fig. 28. Photographs in ordinary transmitted light of cross sections of a TlBr–TlI crystal compressed along the [100] direction. Impact figures were made with a needle within and outside the band; they demonstrate the rotation of the lattice in the kink band. a) The plane of the cross section is perpendicular to the axis of rotation of the lattice; b) the plane of the cross section is parallel to the axis of rotation.

in them are visible in polarized as well as nonpolarized light; (if the slip traces are not visible in nonpolarized light then they cannot be visible in polarized light). The planes of both sections are approximately parallel to the plane of the rhombic dodecahedron. The undeformed region of the crystal is in the lower part of the photographs and the kink band in the upper part. Within the kink band no slip traces are visible but the lattice is rotated, which can be seen by comparing the distribution of the impact figures (made with a needle) within and outside the band. The axis of rotation of the lattice is perpendicular to the section (Fig. 28a) or is in the plane perpendicular to the section (Fig. 28b). The impact figures in Fig. 28b show that the plane of the octahedron replaced the plane of the rhombic dodecahedron in the kink as the result of the rotation of the lattice.

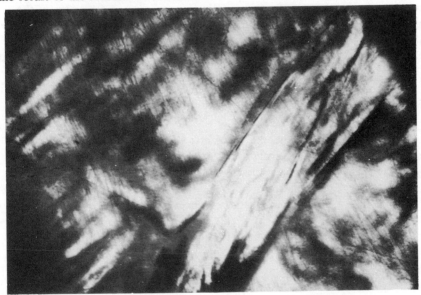

Fig. 29. Photograph in transmitted polarized light of the region of a TlBr–TlI sample in which the kink band was formed without any visible slip traces.

Figure 29 is a photograph of another TlBr– TlI sample taken in polarized light. In this case the kink band was formed also without visible slip traces However the possibility is not excluded that slip does participate in the formation of such regions. Possibly very small stresses which are not visible on the general background of elastic stresses have occurred along the slip traces. The kink band is not a twin interlayer in either sample. This was confirmed in the following way. First the indices of the kink boundary with respect to the lattice of the original crystal were determined on the basis of the impact figures. Then the indices of the plane common to the two mutually rotated lattices (their plane of symmetry) were determined from direct and inverted Laue diagrams of the kink boundary; the Laue diagrams also contained reflections of the original crystal and the kink band. The indices of the plane of symmetry of the section of the sample shown in Fig. 28 are (331). This plane makes an angle of 2° with the real kink boundary. The indices of the plane of symmetry of two contiguous lattices in the sample shown in Fig. 29 are (117) and the plane makes an angle of 5° to the kink boundary. This means that this kink band is not a twin interlayer.

Contrary to the phenomena mentioned previously, E. V. Klontsova et al. [38], observed that in CsI crystals the lattice in the kink is rotated with respect to the original crystal according to the law of noncoherent twinning, with a twinning plane (113), at the points of maximum deformation of the sample (corresponding to points B and D in Fig. 9). The structure of samples with noncoherent twins is given in Fig. 4 (diagram by Brilliantov and Obreimov). However, our data, and also the results obtained by Stepanov and Donskoi [8], show that the position of adjacent rotated parts of the crystal is not always symmetrical. More characteristic of the structure of kinks is the fact that the boundary between the rotated regions is perpendicular to the slip plane.

The boundary between adjacent rotated lattices is symmetrical with respect to them and consequently has a lower potential energy than an asymmetrical boundary. However, under real experimental conditions many factors can prevent the boundary from being symmetrical, factors such as imperfections of the crystal structure, the presence of impurities (particularly probable in crystals of a solid solution of TlBr–TlI), and particularly the distribution of stresses in the sample subjected to deformation.

The regularity of the position of the kink band with respect to slip elements indicates that the formation of kinks is related to the process of slip. This regularity indicates that kink formation is a property correlated to the structure of the crystal.

We studied the process of formation of kinks by using a movie camera with a speed of 5000 frames per second [49]. We used compression by impact to synchronize the first frame with the beginning of deformation. The use of polarized light made it possible to follow the redistribution of stresses (in CsI crystals) and the reorientation of the lattice (in naphthalene crystals) during compression. The orientation of the samples ensured slip in only one of the possible slip systems and it is in this system that an intense plastic deformation took place before as well as during the formation of kinks.

A sharp redistribution of stresses immediately precedes the formation of kinks. Slip in the initial stage of formation of the kink band always occurs in the direction opposite that of the preceding plastic deformation. After this,

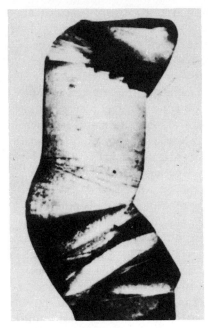

reoriented wedge-shaped regions are formed, and in each successive series of wedges the slip occurs in the opposite direction. If a second interlayer is in the opposite direction.

This change in the direction of slip is connected in each case to the change in the sign of tangential stresses acting on the slip plane in the direction of slip. This occurs because: a) the axis of stress deviates from the axis of the sample due to plastic distortion of the ends of the sample; b) the relative displacement of parts of the sample separated by a kink band due to the localization of slip in this band; c) the rate of rotation of the kink boundary lags behind the rate of plastic deformation within the kink band.

Fig. 30. Naphthalene sample in which kink bands were created as the result of compression (Perekalina, Regel', Dubov).

The study of these phenomena showed that the occurrence of kinks is due to the loss of plastic and not elastic stability of the sample. The determining factor is the redistribution of stresses before the formation of the kink band and during the reorientation of the lattice.

The participation of slip in the rotation of the lattice is considered by Indenbom in "Description of the simplest phenomena of plastic deformation from the viewpoint of dislocation theory," in this collection.

To conclude this section of the article we must note that considerable deformation can result from kinks. This is particularly clear in the case of naphthalene crystals deformed up to 75% by compression. In this case deformation was due exclusively to the formation of kinks [39]. Figure 30 shows such a naphthalene sample.

COMPARISON OF DIFFERENT EFFECTS OF PLASTIC DEFORMATION RESULTING IN IRREGULAR ROTATIONS OF THE CRYSTAL LATTICE

The investigations previously described show that plastic deformation of crystals resulting in rotation of the lattice can be divided into a number of types which differ in their external shapes.

In fact, Brilliantov and Obreimov write that in the case of noncoherent twins the following process is characteristic: "The junction plane of two parts of the crystal is not a coherent crystallographic plane (i.e., it can only be a plane with very high indices)" and "the angle of rotation is not constant, i.e., we can produce crystals with parts rotated at very different angles, from a few minutes to a few degrees." Both of these phenomena are characteristic of kink bands and one can consider that twins along noncoherent planes and kink bands are essentially the same phenomenon. Orowan also considered noncoherent twins and kinks to be the same thing. In confirmation of this we shall note a number of details characteristic of both kink bands and Brilliantov and Obreimov's "twins." Brilliantov and Obreimov write: "The edge of the obtuse angle between two twins is never a straight line but always somewhat curved . . . this means that the boundary consists not of some strongly bent crystal, or a new phase similar in form, or a translational crystal, but consists of several different crystals." The kink band consists also of a number of regions rotated with respect to each other.

The phenomenon of layering observed in rock salt crystals by Stepanov and Donskoi can also be included among the phenomena similar to noncoherent twinning and kink formation. Stepanov and Donskoi consider that the layers occur as the result of reorientation of large regions of the crystal limited by noncrystallographic planes. These authors observed layering of rock salt crystals under conditions similar to those under which kinks are formed in metals (e.g., iron [13]), namely as the result of elongation in the direction of slip at high temperature. This gives us another reason for assuming that layering is one of the types of asymmetric reorientations of the lattice.

Barrett's deformation bands and Jillson's accommodation bands also consist of irregularly rotated regions, and on this basis these phenomena can be considered related to kinks. All these phenomena have the characteristic that they are related to rotation of the material around a given crystallographic direction, i.e., a direction in the slip plane and perpendicular to the direction of slip. The difference between these types of deformation is only in the external shapes of the reoriented regions due to the different conditions of deformation.

Lately there has been a tendency to call any deformation related to rotation of the lattice a kink formation. However, in spite of the relationship between all the phenomena described, their formation could be based on different atomic processes. From the viewpoint of the dislocation theory, the formation of kink bands begins (Hess and Barrett's hypothesis) with the nucleation of dislocations within the crystal and within the limits of the future kink band. Then these dislocations diverge and form the boundaries of the band. As to the mechanism of the formation of deformation bands, there is the Mott hypothesis [40] (which is taken up in detail by Indenbom in this collection).

According to Mott, during the formation of deformation bands dislocations approach the bands from outside the bands.

Bands formed by secondary slip are somewhat different from the phenomena considered so far. These bands occur in crystals which have several systems of slip planes. In these bands the lattice is also rotated with respect to the original crystal, although this rotation should not be considered as an independent phenomenon. In this case rotation is induced by localization of slip in a given region along a system of planes different from that which acted in the rest of the crystal. The other difference is that the formation of bands of secondary slip is apparently accompanied by a rotation of the slip planes around the normal to the slip planes and not around the direction in the slip plane and perpendicular to the direction of slip. This type of bands have not yet been thoroughly investigated.

In a recent article by Jaoul, Bricot, and Lacombe [47] it was proposed to separate irregular deformation into six types:

1) bands of complex slip;

2) bands of secondary slip;

3) compensated kinks within which there are traces of secondary slip;

4) thin kinks occurring at the beginning of deformation in which the rotation of the lattice is rapidly blocked;

5) wide kinks formed at a somewhat later stage than thin kinks; the rotation of the lattice in this case is not blocked by the occurrence of transverse slips;

6) microkinks formed simultaneously with wide kinks and differing from them by the fact that slips developed outside the microkinks.

Bands of type 1, 2, 3, and 6 are related to geometric conditions of deformation. They compensate the total rotation of the lattice of the sample occurring as the result of slip along the principal system of slip planes. Bands of type 4 and 5 occur as the result of irregularity in the crystal structure.

In comparing the results of all the investigations discussed here we shall briefly enumerate the principal data concerning deformation resulting in reorientations of the lattice. Irregularly reoriented regions occur when conditions of deformation are such that bending stresses exist together with compression or tensile stresses. In metals the presence of local defects also favors the occurrence of regions with a reoriented lattice. The occurrence of such regions as the result of compression is favored by rapid deformation, which impedes slipping or twinning, while in the case of elongation it is favored by slowly applied stress, which prevents premature rupture.

Under the described conditions bands consisting of several regions rotated with respect to each other are formed in the crystal; the number of these regions and the total angle of rotation of the lattice within the band increase with the degree of deformation. The boundaries of these bands are approximately perpendicular to the slip planes located in the undeformed part of the crystal. However, in different cases these boundaries (planes) have different indices, which are sometimes very high. Bands with a rotated lattice most frequently have very sharp linear boundaries.

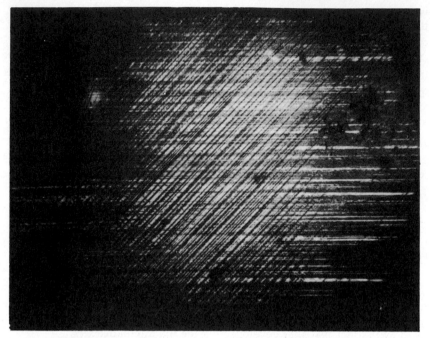

Fig. 31. Surface of the side of an LiF crystal compressed in the vertical direction. The surface has been etched with 3% H_2O_2. The horizontal rows of etch pits (screw dislocation) correspond to the (101) $[10\bar{1}]$ slip system while the oblique rows (edge dislocations0 correspond to the $(0\bar{1}1)$ $[011]$ slip system (Indenbom, Urusovskaya).

At the present time the main subject of discussion is the mechanism of the reorientation of the lattice: does the lattice rotate as the result of deformation by slip, or is the rotation of the lattice analogous to twinning? Only experiments can answer these questions. One such recent experiment concerned noncoherent twinning in NaCl and LiF crystals (Indenbom and Urusovskaya). Using x-ray and optical methods of analysis and selective etching, these authors showed that the formation of noncoherent twins is the result of the action of different slip systems in different parts of the crystal. Figure 31 represents an etched surface of the side of an LiF crystal subjected to compression in the vertical direction. The oblique rows of etch pits correspond to the edge dislocations within the band which extend to the surface, while the horizontal rows correspond to screw dislocations which extend to the surface. This photograph shows that different slip systems act within the band and within the surrounding material.

From this we can conclude that the bands of secondary slip investigated by Honeycombe are nothing else but noncoherent twins.

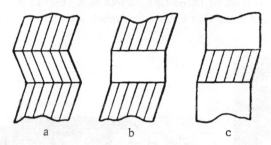

Fig. 32. Diagram of the reorientation of the lattice in crystals by slip. a) Formation of noncoherent twins (slip in neighboring regions in different slip systems; b) formation of deformation bands (slip outside the band); c) formation of slip bands (slip localized within the band (Indenbom, Urusovskaya).

Thus the experimental and theoretical data existing at the present time make it possible to propose three schemes for the reorientation of the lattice in localized regions of the crystal by slip deformation. These schemes are shown in Fig. 32.

FORMATION OF BANDS WITH A ROTATED LATTICE IN POLYCRYSTALLINE METALS AS THE RESULT OF DEFORMATION

An irregular distribution of stress is a condition for the occurrence of asymmetrically oriented regions in a crystal. Therefore we can assume that this phenomenon must play an important role in the plastic deformation of polycrystals since this condition always exists in polycrystalline samples. In fact, in the deformation of samples composed of numerous grains there must always exist an interaction between the grains. Even under conditions of homogeneous elongation this interaction may induce bending moments at the grain boundaries which are capable of causing the formation of kinks or other regions with a rotated lattice within separate grains. This statement is confirmed by the metallographic and x-ray investigations of deformed polycrystals which have been published. In these publications it is reported that very often kink bands, deformation bands, or bands of secondary slip occur as the result of plastic deformation of polycrystals. The formation of kink bands within grains of deformed samples has been reported most often as the result of deformation by elongation, fatigue tests, and creep tests at high temperature. The occurrence of kink bands as the result of fatigue tests was described by Forsyth [41] who subjected polycrystalline aluminum samples to static and cyclic stresses. It must be noted that Forsyth used samples of different degrees of purity and found that kinks occur much more easily in impure samples.

Fig. 33. Kink bands in samples of large grained pure aluminum subjected to creep tests at 600°C (Chaudhury, Grant, Norton).

The occurrence of bands with a disoriented lattice was also observed in a number of investigations of creep in metals [42, 43, 17]. A. Gervais, J. Norton, and N. Grant [42], and H. Chang and N. Grant [43] describe the formation of bands with an asymmetrically rotated lattice in aluminum subjected to very slow elongation at 400-600°C. Figure 33a shows kink bands in a sample of large grained aluminum after a creep test at 600°C. The figure shows three grains, which constituted the sample. In the middle grain one can see the deformation of the slip traces. Figure 33b shows a region of the same sample at a higher magnification.

A. Chaudhury, N. Grant, and J. Norton [17] investigated the formation of kinks in polycrystalline magnesium subjected to creep tests.

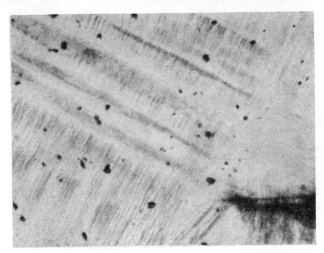

Fig. 34. Kink bands in one of the grains of polycrystalline aluminum subjected to elongation (Honeycombe).

In conclusion we show a photograph obtained by Honeycombe [30] (Fig. 34) which represents one of the grains of a sample of polycrystalline aluminum which was subjected to ordinary elongation. Several kink bands are visible within the grain.

Thus the examples given here clearly illustrate the possibility of the formation of bands with a rotated lattice in deformed polycrystalline metals.

LITERATURE CITED

1. O. Lehman, "Mikrokristallographische Untersuchungen," Z. Kristallogr. 10, 321 (1895).
2. O. Mügge, "Über Translation und verwandte Erscheinungen in Kristallen," Neues Jahrb. Mineral. 1, 71-158 (1898).
3. R. Brauns, "Über die Bedeutung der Morphotropie für Mineralogie," Neues Jahrb. Mineral. 1, 113 (1889).
4. N. A. Brilliantov and I. V. Obreimov, "Plastic deformation of rock salt. III," Zhur. éksp. teor. fiz. 5, No. 3-4, 330-339 (1935); "Plastic deformation. IV," Zhur. éksp. teor. fiz. 7, No. 8, 878-886 (1937).
5. V. I. Startsev, "Rotation of the crystal lattice resulting from plastic deformation," Zhur éksp. teor. fiz. 10, No. 6, 703 (1940); "Transition zones

in plastically deformed rock salt crystals," Doklady Akad. Nauk SSSR 30, No. 2, 124-125 (1941).

6. M. V. Klassen-Neklyudova, "Mechanical twinning of rock salt crystals according to the spinel law," Zhur. éksp. teor. fiz. 12, No. 9, 349-357 (1942).

7. A. V. Stepanov, "Causes of premature rupture," Izv. AN SSSR, ser. fiz. No. 4-5, 797 (1937).

8. A. V. Stepanov and A. V. Donskoi, "New mechanism of plastic deformation of crystals," Zhur. tekh. fiz. 24, 2, 161 (1954).

9. N. J. Selyakov, "Mechanism der Plastizität," Z. Kristallogr. 83, 426-447 (1932).

10. E. Orowan, "A type of plastic deformation new in metals," Nature 149, No. 3788, 643-644 (1942).

11. J. Hess and C. Barrett, "Structure and nature of kink-bands in zinc," Trans. AIMME 185, 599-604 (1949).

12. J. Gilman, "Mechanism of ortho-kink-band formation in compressed zinc monocrystals," J. Metals 6, Sec., 2, No. 5, 621-629 (1954).

13. A. Holden and F. Kunz, "Dimension and orientation effects in the yielding of carburized iron sheet crystals," Acta metallurg. 1, No. 5, 495-502 (1953).

14. A. Jackson and B. Chalmers, "Influence of striations on the plastic deformation of single crystals of tin," Canad. J. Phys. 31, No. 6, 1017-1018 (1953).

15. J. Washburn and E. Parker, "Kinking in zinc single crystal tension specimens," J. Metals 4, No. 10, 1076-1078 (1952).

16. J. Gilman and T. Read, "Bend-plane phenomena in the deformation of zinc monocrystals," J. Metals 5, No. 1, 49-55 (1953).

17. A. Chaudhury, N. Grant, and J. Norton, "Metallographic observation on the deformation of the high-purity magnesium in creep at 500°F," J. Metals 5, No. 5, 71-76 (1953).

18. A. T. Churchman, "The yield phenomena, kink-band and geometric softening in titanum crystals," Acta metallurg. 3, No. 1, 22-29 (1955).

19. V. R. Regel' and V. G. Goborkov, "Effect of temperature T and deformation rate v on the critical shear stress σ_{cr} in zinc monocrystals," Kristallografiya 3, No. 1 (1958).

20. L. Pfeil, "Effect of cold work on structure and changes produced by subsequent annealing," Carnegie School. Mem. Iron and Steel Inst. 16, 153-210 (1927).

21. C. S. Barrett, "Structure of iron after compression," Trans. AIMME 135, 246-322 (1939).

22. J. Herenguel and P. Lelong, Compt. rend. 233, 53-55 (1951).

23. J. Friedel, "The mechanism of work-hardening and slip-band formation," Proc. Roy. Soc. A, 242, No. 1229, 147-159 (1957).

24. D. Jillson, "An experimental survey of the formation and annealing processes in zinc," Trans. AIMME 188, 1009-1017 (1950).

25. A. Moore, "Accommodation kinking, associated with the twinning of zinc," Proc. Phys. Soc. B, 12, 956-958 (1952).

26. A. Moore, "Twinning and accommodation kinking in zinc," Acta metallurg. 3, No. 2, 163-169 (1955).

27. P. Pratt and S. Pugh, "Twin accommodation in zinc," J. Inst. Metals. 80, 653-658 (1952).

28. C. Barrett, "Recrystallization texture of aluminum after compression," Trans. AIMME 137, 128-144 (1940).

29. E. Calnan, "Laue asterism and deformation bands," Acta crystallogr. 5, 557-564 (1952).

30. R. Honeycombe, "Inhomogeneities in the plastic deformation of metal crystals," J. Inst. Metals 80, 45-56 (1951-1952).

31. H. Wilman, "Rotational slip— a new deformation process in crystals," Nature 165, No. 4191, 321 (1950).

32. A. B. Zemtsov, M. V. Klassen-Neklyudova, and A. A. Urosovskaya, "Complex phenomena of the plastic deformation of monocrystals," Doklady Akad. Nauk SSSR 91, No. 4, 913-816 (1953).

33. M. V. Klassen-Neklyudova and A. A. Urusovskaya, "Effect of irregular distribution of stress on the mechanism of plastic deformation of thallium and cesium halides," Kristallografiya 1, No. 4 (1956).

34. M. V. Klassen-Neklyudova and A. A. Urosovskaya, "The structure of kink bands in thallium halide crystals," Kristallografiya 1, No. 5, 564-571 (1956).

35. M. V. Klassen-Neklyudova, A. A. Urosovskaya, "Through impact and pressure figures in cubic halide crystals," Trudy In-ta kristallogr. AN SSSR 11, 146-151 (1955).

36. A. A. Urusovskaya, "Impact and pressure figures in crystals of hexagonal metals," Trudy In-ta kristallogr. AN SSSR 11, 140-145 (1955).

37. A. A. Urosovskaya, "Figures of plastic deformation in TlBr— TlI, CsI, and CsBr crystals," Trudy In-ta kristallogr. AN SSSR 12, (1956).

38. E. V. Kolontsova, I. V. Telegina, and G. M. Plavink, "Structure of kink bands in some ionic crystals," Kristallografiya 1, No. 4 (1956).

39. Z. B. Perekalina, V. R. Regel', and G. A. Dubov, "Some results of compression tests of naphthalene monocrystals," Kristallografiya 3, No. 1 (1958).

40. N. F. Mott, "The mechanical properties of metals," Proc. Phys. Soc. B, 64, 729 (1951).

41. P. Forsyth, "Some metallographic observations on the fatigue of metals," J. Inst. Metals. 80, 181 (1951).

42. A. Gervais, J. Norton, and N. Grant, "Kink-band formation in high-purity aluminum during creep at high temperatures," J. Metals 5, Sec. 2, No. 11, 1487-1492 (1953).

43. H. Chang and N. Grant, "Mechanism of creep deformation in high-purity aluminum at high temperatures," J. Inst. Metals 82, No. 6, 229-235 (1954).

44. A. Anderson, D. Jillson, and B. Dunbar, "Deformation mechanisms in alpha-Ti," J. Metals 5, Sec. 2, No. 9, 1191-1197 (1953).

45. W. Berg, "Mechanical twinning in bismuth crystals," Nature 134, 143 (1934).

46. J. Diehl, "Zugverformung von Kupfer-Einkristallen. I. Verfestigungskurven und Oberflächenerscheinungen," Z. Metallkunde 47, H. 5, 331-343 (1956).

47. B. Jaoul, J. Bricot, and P. Lacombe, "Les bands de deformation et les pliages dans les monocristaux d'aluminium," Rev. metallurg. 54, No. 10, 756-768 (1957).

48. M. V. Klassen-Neklyudova and A. A. Urusovskaya, "Plastic deformation of rock at high temperature," Kristallografiya 5, No. 5, 744-748 (1960).

49. M. V. Klassen-Neklyudova, G. E. Tomilovskii, and M. A. Chernysheva, "Process of formation of kinks," Kristallografiya 5, No. 4, 464-469 (1960).

50. V. I. Startsev and V. M. Kosevich, "Surface relief on habit planes of bismuth, antimony, and zinc created by twin interlayers," Fizika metallov i metallovedenie 2, 320 (1956).

51. V. I. Startsev and F. F. Lavrent'ev, "X-ray investigation of accommodation bands in zinc resulting from twinning," Kristallografiya 3, 329-333 (1958).

V. L. INDENBOM

DESCRIPTION OF THE SIMPLEST
PHENOMENA OF PLASTIC DEFORMATION
FROM THE VIEWPOINT OF DISLOCATION
THEORY

The basic assumptions of the dislocation theory have been confirmed in recent years. The electron microscope has made it possible to observe directly atomic layers in crystals of platinum and copper phthallocyanides and to observe regions in which the relative positions of atomic planes indicate the presence of dislocations (Menter [1]). In some cases (Hashimoto and Uyeda [2], Pashley, Menter, and Basset [3]) it was possible to detect the dislocation by comparing the moiré patterns of deformed and undeformed crystal lattices, as proposed by Shubnikov [4, 5].

In electron microscope investigations of a thin aluminum foil (Hirsh, Horne, and Whelan [6]) and of stainless steel (Bollmann [7], Whelan, Hirsh, Horne, and Bollmann [8]) the dislocations were located by the increased intensity of electron scattering. Motion pictures were taken which showed the distribution of dislocations and their interaction, their movement under stress, etc.

The methods of selective etching and segregation of impurities became widely used in the study of dislocations when it became known that the method of metallographic investigation of crystal defects could be improved to such an extent that it was possible to distinguish separate dislocation lines (Vogel [9, 10], Amelinckx [11, 12], Dash [13], Indenbom and Tomilovskii [14]). A more complete review of modern methods of investigating dislocations (including x-ray diffraction, optical, and others) can be found in [97, 96, 99, 100, 101].

As has been shown [15-17], the accumulated experimental data indicate that dislocations really exist and move in crystals and have properties of elementary lattice defects and sources of internal stress predicted by the theory. Since the assembly of dislocations in rows and nets leads to the breaking of the crystal into blocks and the motion of dislocations leads to plastic deformation of crystals, it is necessary to take into account the role of dislocations in the study of the structure of real crystals as well as in the study of the mechanism of plastic deformation. Obviously it is still a long way from the dis-

105

location hypothesis to the creation of the dislocation theory of the mechanical properties of crystals, but even at the present stage of development it is necessary to attempt to use the established experimental data relative to the properties of separate dislocations in analyzing the mechanism of the so-called elementary acts of plastic deformation, which are apparently connected with the formation and displacement of hundreds and thousands of dislocations.

1. TRANSLATIONAL SLIP RESULTING FROM THE MOTION OF DISLOCATIONS

After Obreimov and Shubnikov [18] (who analyzed the plastic deformation of NaCl crystals in polarized light) showed that slip develops progressively, it became clear that translational slip cannot be reduced to simultaneous slipping of crystalline blocks and must necessarily be accompanied by the formation of an atomic nonius at the boundary of the slipping parts of the crystal. Considering this assumption obvious (if only translational slip occurs), Obreimov and Brilliantov [19] expressed doubts of the existence of translational slip as such. At the same time Orowan, Polanyi, and Taylor took the opposite view and created the foundation of the modern theory of dislocation, assuming that translational slips are real and that they can be spread by the motion of atomic noniuses— dislocations.

Orowan [20] showed that if a limited region of one atomic plane slips along the neighboring plane for a distance equal to the parameter of the translation of the lattice, then the boundary of this region forms a closed ring of dislocations. Further, slips can extend along the slip plane by gradual widening of the region of local slip, which is equivalent to the displacement of dislocations in the slip plane. Polanyi [21] investigated a read edge dislocation and noted that the displacement of the dislocation leads to slip by an amount equal to the parameter of the lattice. He attempted to calculate the stress necessary for the displacement of dislocations. Taylor [22] gave a scheme of the distribution of atoms in an edge dislocation which differs very little from modern concepts. Contrary to Orowan, Taylor considers that the atomic planes next to the dislocation do not remain straight but continue to bend around the edge of the extra half plane which forms the dislocation line. Taylor showed that translational slip for a length equal to the parameter of the lattice can occur due to the motion of positive as well as negative edge dislocations.

The assumption that slip propagages along the slip plane was later confirmed by those authors who did not agree with the dislocation theory of plasticity (Kontorova and Frenkel' [23], Klassen-Neklyudova and Kontorova [24], Stepanov [25], Garber, Obreimov, and Polyakov [26], Bochvar and Preobrazhenskaya [27]). It must be noted, however, that the gradual propaga-

tion of slip has necessarily a dislocation character insofar as the boundary of
the region of local slip represents a dislocation line (as has been determined
by Read [28]). In particular, the sections of the boundary perpendicular to
the direction of slip form edge dislocations (Fig. 1a) while the sections of the
boundary parallel to the direction of slip form screw dislocations (Fig. 1b).

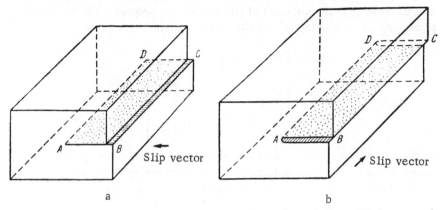

Fig. 1. Dislocation as a boundary of the region of local slip. Slip has passed
through the ABCD area. AD) Dislocation; a) edge dislocation (perpendicular
to the slip vector); b) screw dislocation (parallel to the slip vector).

Calculations made on the basis of the Frenkel'-Kontorova model [23,29],
the Peierls-Nabarro model [30-34], and the Born model [35] show that the '
"width" of a single dislocation represents several lattice parameters, i.e., the
effective width of the transitional region between the regions along which the
magnitude of slip differs by one lattice parameter. It follows then that any
irregular slip is of a discrete nature and can be separated into single dislocation lines (if the gradient of the magnitude of slip does not exceed the value
of one lattice parameter over a length of the order of the width of the dislocation line). Thus, for example, a slip which increases regularly along the direction of slip corresponds to a series of equidistant edge dislocations. The
slip which increases regularly in the direction perpendicular to the direction
of slip corresponds to a series of equidistant screw dislocations.

SLIP PLANES AND MACROSCOPIC STRESSES

Because of the irregularity of slip along the slip plane the displacements
and deformations do not coincide on each side of the slip plane. If N dislocations pass through some point of the slip plane then a slip vector **b** (Burgers
vector) corresponds to each of these dislocations and the displacement vector
u, on each side of the slip plane will differ by N**b**:

$$[u] = Nb \tag{1}$$

(the bracket stands for a jump of the function).

One can calculate macroscopic deformation on the assumption that N is continuously dependent on the coordinates. Let u, v, and w be the components of the displacement vector along the X, Y, and Z axes, and let the X axis be parallel to the direction of slip and Z axis perpendicular to the slip plane. By differentiating relationship (1) we obtain:

$$\left.\begin{aligned}
\left[\frac{\partial u}{\partial x}\right] &= [e_x] = b\,\frac{\partial N}{\partial x}\,, \\[4pt]
\left[\frac{\partial u}{\partial y}\right] &= [\gamma_{xy}] = b\,\frac{\partial N}{\partial y}\,, \\[4pt]
\left[\frac{\partial v}{\partial x}\right] &= \left[\frac{\partial v}{\partial y}\right] = 0.
\end{aligned}\right\} \tag{2}$$

By Hooke's law we can calculate the jump of macroscopic stresses by the jumps of the macroscopic deformations. Stresses acting on the slip plane remain, of course, continuous. For an elastic isotropic crystal we have:

$$[\sigma_x] = \frac{2G}{1-\nu}[e_x] = \frac{2bG}{1-\nu}\,\frac{\partial N}{\partial x}\,, \tag{3}$$

$$[\sigma_y] = \nu\,[\sigma_x], \tag{4}$$

$$[\tau_{xy}] = G\,[\gamma_{xy}] = bG\,\frac{\partial N}{\partial y}\,, \tag{5}$$

where G is the shear modulus and ν the Poisson coefficient.

Edge dislocations [N = N(x)] induce the jump of normal stresses acting along the direction of slip and along the dislocation line. Screw dislocations [N = N (y)] induce a jump of tangential stresses acting along the direction of slip on the plane perpendicular to the slip plane and parallel to the dislocation lines.

In the particular case of regular distribution of dislocations in the slip plane, relationships (2) to (5) and their generalization for anisotropic crystals was considered by Nye [36], and Indenbom and Tomilovskii [14]. For a row of equidistant edge dislocations of the same sign, we have:

$$N = \frac{x}{h} + \text{const}, \tag{6}$$

where h is the distance between the lines; for a row of equidistant screw dislocations, we have:

$$N = \frac{y}{h} + \text{const}. \tag{6'}$$

Macroscopic stresses surrounding the slip plane were measured for the first time by Obreimov and Shubnikov [18]. They investigated NaCl crystals by

the polarization-optical method and showed that slip planes induce jumps of
macroscopic stresses which are visible when viewed in the directions perpendicular and parallel to the slip plane. In other words, Obreimov and Shubnikov
observed (in our notations) the jump of stress σ_x as well as the jump of stress
σ_y. The maximum stress was 9.5 kg / mm², which according to formulas (3)
and (6) corresponds to a distance of the order of tenths of a micron between
dislocations when b = 2.81 A · √2 = 3.97 A. The authors noted an elastic
bend of regions of the crystal located between neighboring slip bands and attempted to construct the line of main stresses but did not take into account
the anisotropy of photoelasticity of the crystal.

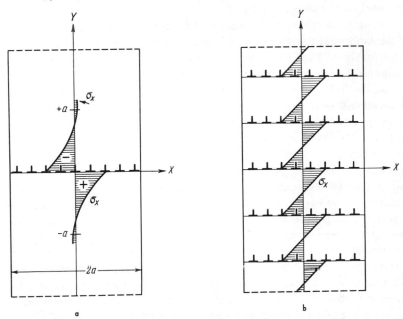

Fig. 2. Calculated distribution of macrostresses acting parallel to
the direction of slip induced by edge dislocations. The stresses are
given for the plane of the sample. a) A row of edge dislocations
(single slip band); b) a series of equidistant slip planes with the
same density of edge dislocations.

Nye [36] made a detailed investigation of the distribution of stresses surrounding the slip bands in AgCl crystals. Like Obreimov and Shubnikov, he
found that as a rule the jumps of stresses for a given system of slip planes are
of the same sign. Consequently the dislocations (excess dislocations) in these
planes have the same sign. From this result he concluded that slip ordinarily

begins at the surface of one side of the crystal.* Nye compared the distribution of stresses found by optical methods with the results calculated by formula (3) and (5). Figure 2a shows the calculated distribution of stress σ_x for a crystalline sample with a single row of edge dislocations. From the condition that the stress acting on the free surface of the sample = 0 it follows that the stresses decrease progressively away from the slip plane and become almost zero at a distance of the order of the width of the sample. If the neighboring active slip planes are located at a distance which is small with respect to the cross sectional dimensions of the sample (Fig. 2b) the region of the crystal between the bands must undergo a homogeneous elastic bend. This was the usual result observed by Nye.†

In order to calculate the density of dislocations Nye measured the jump of the tangential stresses on one of the slip planes and obtained a value of approximately 0.25 kg/mm^2. According to Eqs. (5) and (6) this corresponds to a distance between dislocations of h = 1.6 μ when b = 5.54 A/$\sqrt{2}$ = 3.92 A.

Recently Indenbom and Tomilovskii [14] succeeded in showing that the dislocation density in slip planes actually corresponds to the value calculated by the jump of macroscopic stresses. The sample was a corundum plate (Fig. 3). The distance between the etch pits in the slip lines and the difference between the normal stresses acting in the direction of slip on both sides of the slip plane were measured directly. At the same time this difference was calculated according to a formula similar to formulas (3) and (6) on the assumption that each etch pit corresponds to the end of a single edge dislocation with a Burgers vector b equal to 8.3 A, this value being equal to the minimum translation vector along the direction of slip. The discrepancy between the theoretical and experimental result did not exceed 20%. Later the same authors discovered in the slip bands of corundum [102] a microstructure of the stress field related to the dislocational structure of slip bands. By adapting an electron microscope they succeeded in observing directly the stress fields around 90° (edge) and 60° dislocations in silicon [103].

Thus in the simplest case of regular distribution of dislocations the translational slip in the active slip planes must be regarded not simply as slipping

---

*See also the qualitative observations of Stepanov, Mil'kamanovich, Melankholin, and Regel' [37].

† Further investigation of the AgCl crystal by Kochnov and Shaskol'skaya [38] showed that the slip lines, which can easily be seen by the steps on the surfaces of the sample, sometimes do not coincide with bands of double refraction. Therefore the authors investigated slip bands containing screw dislocations along the direction of slip. In this case macroscopic stresses surrounding the slip bands should not induce a visible double refraction in the same way as screw dislocations, observed along the axis by Indenbom and Tomilovskii [39], do not create double refraction.

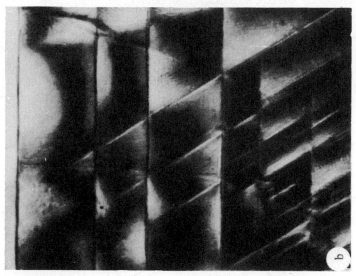

Fig. 3. Slip bands in a corundum crystal. The jump of stresses on each side of the slip planes corresponds to the density of dislocations calculated by the etch pits. a) Revealed by etching method; b) revealed by polarization-optical method.

of parallel planes of the crystal by a translation vector but as slip with elastic bending of crystalline interlayers located between neighboring slip bands.*

ACCUMULATIONS OF DISLOCATIONS AT BARRIERS

The motion of dislocations in the slip plane may be stopped at some defect in the crystal, at the block boundaries, or at the surfaces of the crystal. Cottrell [42] showed that a row of n edge dislocations of the same sign which encounters a barrier in the slip plane (Fig. 4a) exercises a strong pressure on this barrier. The head dislocation, which must restrain all the following dislocations, creates a stress on the barrier which is n times greater than the applied stress.

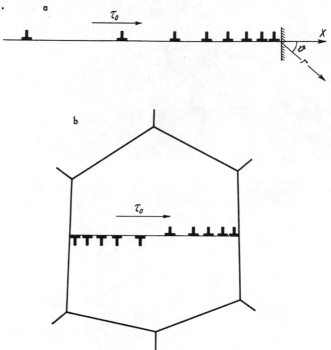

Fig. 4. Accumulation of dislocations in slip planes. a) A row of edge dislocations created by a barrier; b) a single slip band impeded at the boundaries of a crystal grain in terms of dislocations.

Eshelby, Frank, and Nabarro [43] calculated the position of the dislocations, taking into account that each dislocation creates in its own slip plane a tangential stress which decreases in inverse proportion to the distance from

---

* The scheme of "slip with bending" was first proposed by Mügge [40] and was discussed in detail by Polanyi [41].

the dislocation line, and that the dislocation can be in equilibrium only in a place where the total tangential stress (from other dislocations and applied forces) is equal to zero. It turned out that for a homogeneous exterior field the problem is equivalent to finding conditions under which wave functions of hydrogen atoms are zero.

In the limit case of a large number of dislocations, the extent of the accumulation L is proportional to the number of dislocations n and inversely proportional to the external tangential stress, τ_0 acting on the slip plane in the direction of slip

$$L = \frac{nGb}{\pi \tau_0 K}, \tag{7}$$

where b is Burgers vector, and K is equal to unity for screw dislocations and 1-ν for edge dislocations. The distance between the head dislocation and the following dislocation is equal to

$$l = 0.92 \frac{L}{n^2}. \tag{8}$$

At a distance x in front of the head dislocation the tangential stress differs very little from the stress created by one of these dislocations when x $\ll l$. When $l \ll$ x $\ll$ L, then

$$\tau(x) \approx \tau_0 \sqrt{L/x}, \tag{9}$$

but if x $\gg$ L then the stresses added to τ_0 are equal to those which would be created by one dislocation with a Burgers vector of nb situated in the center of gravity of the accumulation.

Leibfried [44] assumed that the distribution of dislocations can be considered continuous if their number is very large. Then the solution of the problem concerning the distribution of dislocations in a field of given external forces is reduced to the solution of integral equations of the type

$$\tau_0 + \frac{Gb}{2\pi K} \int \frac{D(x')\,dx'}{x - x'} = 0, \tag{10}$$

where D is dislocation density and the integral as a whole represents the stresses induced by dislocations. In particular, for a slip plane which has equal numbers of positive and negative dislocations on both sides for a distance 2a the solution of Eq. (10) gives the following expression for the dislocation density:

$$D(x) = -\frac{2\tau_0 K}{Gb} \cdot \frac{x}{\sqrt{a^2 - x^2}}. \tag{11}$$

The total number of dislocations of a given sign can be found by integrating the dislocation density

$$n = \int_{-a}^{0} D\,(x)\,dx = \frac{2\tau_0 aK}{Gb}, \tag{12}$$

which is equivalent to Eq. (7) when L = $2a/\pi$. Figure 4b illustrates this solution in the case of a single slip line stopped at the boundaries of a crystal grain.

Accumulations of dislocations were first observed experimentally by the etching method in polycrystalline brass (Jacquet [45]).

Bilby and Entwisle [46] compared the distribution of etch pits with the prediction of the theory [43] and came to the conclusion that each etch pit corresponds to a dislocation end although there are small systematic differences in the distribution of the etch pits and the theoretical distribution of dislocations calculated by Eqs. (7) to (9).

Later, Meleca [47] and Gilman [48] investigated accumulations of dislocations in zinc and noted a qualitative agreement between the experiments and the theory. Meleca, for example, using the length of the accumulation and the applied stress, calculated by formula (7) the number of dislocations in the accumulation and found n = 13.8, while the experiment showed n = 11. Indenbom and Tomilovskii [39] observed accumulations of dislocations at the ends of impeded slips by optical methods. The distribution of stresses corresponded qualitatively to that induced by one macroscopic dislocation with a Burgers vector equal to the sum of Burgers vectors of dislocations accumulated at the end of the slip. From an electron microscope investigation of stainless-steel foil Whelan, Hirsh, Horne, and Bollmann [8] obtained beautiful photographs of accumulations of dislocations at the grain boundaries. The distribution of dislocations was in good qualitative agreement with that shown in Fig. 4. Recently Rozhanskii and Indenbom [104 examined models of accumulations of dislocations formed by successive blocking of dislocations by impurities; they showed that the calculated distribution of dislocations in this model is in better agreement with the experimental data than in the model of freely slipping dislocations [43, 44].

Stroh [49], using the results of Eshelby, Frank, and Nabarro [43], made similar calculations of the field stress at the end of an impeded slip. For stresses in front of an impeded row of edge dislocations, formula (9) can be generalized by use of the notations in Fig. 4a:

$$\left.\begin{aligned} \frac{\sigma_x + \sigma_y}{2\tau_0} &\approx \sqrt{\frac{L}{r}}\sin\frac{\vartheta}{2}\,, \\ \frac{\sigma_x - \sigma_y}{2\tau_0} &\approx \sqrt{\frac{L}{r}}\sin\frac{\vartheta}{2}\left(1 + \cos\frac{\vartheta}{2}\cos\frac{3\vartheta}{2}\right), \\ \frac{\tau_{xy}}{\tau_0} &\approx \sqrt{\frac{L}{r}}\cos\frac{\vartheta}{2}\left(1 - \sin\frac{\vartheta}{2}\sin\frac{3\vartheta}{2}\right). \end{aligned}\right\} \tag{13}$$

The relationships show that the plane with azimuth ϑ passing through the head dislocation is acted upon by normal stress:

$$\sigma \approx \frac{3}{2} \sqrt{\frac{L}{r}} \, \tau_0 \sin \vartheta \cos \frac{\vartheta}{2} . \tag{14}$$

The concentration of stresses induced by a large accumulation of dislocations may turn out to be sufficient to destroy the material. Figuratively speaking, the point of the slip line can split the crystal.

In expression (14) the maximum is reached when $\cos \vartheta = \frac{1}{3}$, i.e., when $\vartheta = 70.5°$. This azimuth corresponds to the most probable direction of the crack in the crystal with an isotropic strength. Obviously, in crystals with well-defined stacking planes the cracks will be different. However, taking into account the role of microcracks in the process of slip formation assumed by Garber, Obreimov, and Polyakov [26], Bochvar and Preobrazhenskaya [27], we must emphasize that expression (14) does not indicate the existence of any normal stresses acting on the slip plane.

From formulas (13) and (14) it follows that for a given external stress τ_0, the normal stresses σ are proportional to the square root of the accumulation length L and, consequently, to the square root of the number of accumulated dislocations n, which is proportional to the dimensions of the crystal grain according to formula (12). Consequently, external stress inducing destruction must be proportional to the square root of the characteristic dimension of the crystal grain, which has been confirmed experimentally [50]. The modern dislocation theory of destruction is given in more detail in [105, 106, 107].

LATENT ENERGY OF THE SLIP PLANES

The energy of macroscopic stresses and microscopic deformations induced by rows of dislocations in the slip plane can be interpreted as the latent energy of this plane. In this sense the slip plane becomes analogous to the surface of separation, with a corresponding surface energy.

Let us assume that active slip planes in a crystal are located at a distance s from each other and contain rows of equidistant edge dislocations.

If the cross sectional dimensions of the crystal are much greater than s then the distribution of macroscopic stresses in each elastically bent layer will be given [on the basis of formula (3) and (6)] by the following expression:

$$\sigma_x = [\sigma_x] \left(\frac{y}{s} - \frac{1}{2} \right) = \frac{2Gb}{(1-\nu)h} \left(\frac{y}{s} - \frac{1}{2} \right) . \tag{15}$$

The elastic energy of macroscopic stresses per square centimeter of each active slip plane is

$$\gamma = \frac{1}{2} \int_0^s e_x \sigma_x \, dy = \frac{Gb^2 s}{3(1-\nu)h^2} . \tag{16}$$

The volume energy density w is independent of the distance between the slip bands:

$$w = \frac{\gamma}{s} = \left(\frac{b}{h}\right)^2 \frac{G}{3(1-\nu)} \, . \tag{17}$$

Taking the typical value, $b = 4$ A, $G = 4 \cdot 10^{11}$ d/cm², $\nu = \frac{1}{3}$ when $h = 1\mu$, it follows that $w \approx 3 \cdot 10^{-4}$ erg/cm³, and when $h = 0.1\,\mu$ we obtain $w \approx 3 \cdot 10^{-2}$ erg/cm³. When $s \sim 1\,\mu$ the surface energy of each slip plane is, respectively, 3 and 320 erg/cm².

    In order to take into account the energy related to the fine structure of the stress field, it is necessary to sum rigorously stresses induced by single dislocations. This summation can be made by a method analogous to that indicated by Cottrell [51] and Indenbom and Tomilovskii [102]. Subtracting the macroscopic stresses, we obtain for a row of equidistant edge dislocations

$$\sigma_x = \frac{Gb}{(1-\nu)h} \left\{ 1 - \left(1 + \frac{y}{2}\frac{\partial}{\partial y}\right) \left(\frac{\operatorname{sh}\dfrac{2\pi y}{h}}{\operatorname{ch}\dfrac{2\pi y}{h} - \cos\dfrac{2\pi x}{h}}\right) \right\} . \tag{18}$$

In particular, for cross sections coinciding with excess dislocation planes $(x = \pm\, nh)$ we have

$$\sigma_x = \frac{Gb}{(1-\nu)h} \left(1 + \frac{\dfrac{\pi y}{h}}{2\operatorname{sh}^2\dfrac{\pi y}{h}} - \operatorname{cth}\frac{\pi y}{h} \right). \tag{18'}$$

The energy related to the fine structure of a stress field can finally be obtained by integrating stresses (18') along the excess planes of dislocations. As is well known, this operation corresponds to the calculation of the energy of formation of dislocations (see, for example, Read [28], chapter 8.4). The energy per unit length of each dislocation is

$$W = -\frac{b}{2} \int_{r_0}^{\infty} \sigma_x \, dy = \frac{Gb^2}{4(1-\nu)}\left(-1 + \frac{1}{\pi}\ln\frac{h}{2\pi r_0}\right), \tag{19}$$

where r_0 is the radius of the "nucleus" of dislocation, which can be assumed to be approximately equal to the length of Burgers vector b.* From this, the fraction of the surface energy of the slip plane determined by the microscopic deformation is

$$\gamma = \frac{W}{h} = \frac{Gb^2}{4(1-\nu)h}\left(-1 + \frac{1}{\pi}\ln\frac{h}{2\pi r_0}\right). \tag{20}$$

---

* The parameter r_0 is chosen in such a way that formulae of the type of (19) give correct values for the total energy of the dislocation, including inelastic distortion in the nucleus of the dislocation.

When G = $4 \cdot 10^{11}$ dynes/cm^2, $\nu = \frac{1}{3}$, b = r_0 = 4 A, we have $\gamma \sim 2$ erg/cm^2 when h $\sim 1 \mu$ and $\gamma \sim 4$ erg/cm^2 when h ~ 0.1.

Thus when the distribution of dislocations in the slip plane is regular and the distance between dislocations is of the order of a micron the surface energy of micro- and macrodeformations is a relatively small value— of the order of a few ergs per square centimeter. When the distance between dislocations in the slip band is of the order to one-tenth of a micron, the surface energy due to macrodeformation reaches hundreds of ergs per square centimeter, exceeding the energy of microdeformation by a large amount. The volume density of the latent energy increases proportionally to the square of the density of moving dislocations in slip bands and not proportionally to the average dislocation density, as is usually assumed.

An even higher density of latent energy is obtained when the distribution of dislocations is irregular. Since the field stresses of single dislocations are additive while the elastic energy is a function of the square of the value of the stress, then n closely situated dislocations have an energy approximately n times greater than the total energy of n single dislocations.

According to the precise calculations by Stroh [49] the energy per unit length of each dislocation of an accumulation composed of n edge dislocation is

$$W = \frac{nGb^2}{4\pi(1-\nu)}\left(\ln\frac{4R}{L} + \frac{1}{2}\right), \qquad (21)$$

where L is the length of the accumulation determined by (7) and R is the characteristic dimension of the crystal.

If the distance between the dislocations in the slip plane is equal to h on the average then the averaged surface energy is

$$\gamma = \frac{W}{h} = \frac{nGb^2}{4\pi(1-\nu)h}\left(\ln\frac{4R}{L} + \frac{1}{2}\right) \qquad (22)$$

and exceeds approximately n times the value found in the case of regular distribution of dislocations calculated by Eq. (20). For example, when n = 100 the surface energy reaches hundreds of ergs per square centimeter even when the average distance between dislocations is of the order of 1 μ.

The volume density of latent energy is proportional to the average dislocation density multiplied by the average power of accumulations.

2. INTERACTION BETWEEN DISLOCATIONS AND FORMATION OF REORIENTED REGIONS OF THE LATTICE

Up to now we have considered the motion of dislocations in isolated slip planes. The barriers to such motion were "extraneous" defects of the lattice, e.g., grain boundaries. However, if one takes into account the interaction between dislocations situated in different slip planes one finds that field stresses created by the moving groups of dislocations themselves may constitute barriers capable of preventing further translational slip and lead to the occurrence of other forms of plastic deformation of the crystal. From the viewpoint of dislocation theory the reorientation of the lattice turns out to be a simple consequence of the fact that each single dislocation induces local deformation of the lattice (see [17], for example). From the known formula for the displacement fields corresponding to linear dislocations (Read [28]) it follows that an edge dislocation induces rotation of the lattice (around an axis parallel to the dislocation line) by an angle

$$\omega_z = -\frac{b}{4\pi}\frac{\cos\vartheta}{r} \tag{23}$$

(we have adopted the polar system of coordinates, r, ϑ, z; the azimuth is measured from the Burgers vector).

At a screw dislocation the lattice is rotated around the directions of the radii by an angle

$$\omega_r = \frac{b}{4\pi r}. \tag{24}$$

Since the rotations of the lattice induced by single dislocations [as given by formulas (23) and (24)] decrease slowly with the distance from the dislocation line, the collective effect of a group of dislocations distributed in the proper manner turns out to be sufficient to induce general reorientation of the lattice in separate regions of the crystal.

INTERACTION BETWEEN EDGE DISLOCATIONS OF THE SAME SIGN IN PARALLEL SLIP PLANES. FORMATION OF ROTATED BLOCKS

The field of tangential stresses (induced by an edge dislocation which act on the slip plane in the direction of slip

$$\tau_{xy} = \frac{Gb}{2\pi(1-\nu)}\frac{\cos\vartheta\cos 2\vartheta}{r} \tag{25}$$

determines the interaction between dislocations located in parallel slip planes. The field stress can be characterized as a rosette of equal tangential stresses, $r = C\cos\vartheta\cos 2\vartheta$ (Fig. 5a), which is well illustrated in the photographs obtained by Bond and Andrus [52], Chernysheva [53], and Tomilovskii [39]. Even clearer illustrations were obtained in the investigation of stresses around sessile edge dislocations in silicon [103]. The black and white petals correspond to the sectors in which the tangential stresses, τ_{xy} have different signs. In hatched sectors, corresponding to the small petals of the rosette, edge dislocations of the same sign are attracted. In sectors corresponding to the large

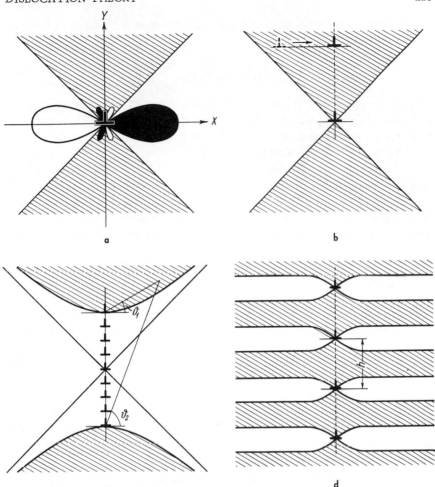

Fig. 5. Scheme of the formation of vertical rows of dislocations. The hatched regions represent the areas of the crystal from which new dislocations were attracted to the row. a) Rosette of equal tangential stresses characterizing the interaction between parallel edge dislocations; b) vertical equilibrium positions of a pair of dislocations of the same sign; c) scheme of the growth of the ends of the row; d) scheme of the increase in density of the row.

petals of the rosette these dislocations are repelled. The equilibrium state is reached when the dislocations are situated "vertically" (Fig. 5b), when their excess planes coincide.

In order for two edge dislocations of the same sign to pass each other in parallel slip planes without stopping in the vertical equilibrium reaction, the external tangential stress must be higher than the maximum stress of interactions of dislocations [28, 51]

$$\tau_m = \tau_{xy}\big|_{\vartheta=\pi/8} = \frac{Gb}{8\pi\,(1-\nu)\,h}, \tag{26}$$

where h = r sin ϑ is the distance between slip planes. If the external tangential stress is less than τ_m then the configuration shown in Fig. 5b occurs. Then these two dislocations can stop the third, fourth, etc., gradually forming a vertical row of dislocations (Fig. 5c).

The hatched areas of Fig. 5c limited by hyperbolae represent areas from which following dislocations can accumulate at the end of the row. The field of stresses can be found by the summation of stresses of single dislocations (25). At distances large with respect to the difference between dislocations the summation can be replaced by integration. This gives (see Frank and Stroh [54])

$$\tau_{xy} = -\frac{Gb}{4\pi\,(1-\nu)\,h}\,(\sin 2\vartheta_1 - \sin 2\vartheta_2), \tag{27}$$

where ϑ_1 and ϑ_2 are azimuths at which the investigated point is observed from the end of the row. It is important to note that stresses at the ends of the row progressively increase as the row builds up, approaching the value

$$\tau_{\max} = \frac{Gb}{4\pi\,(1-\nu)\,h}. \tag{27'}$$

A sufficiently dense row can create such a concentration of stresses that it exceeds the critical shear stress for an ideal crystal, and thus the density of the row begins to increase spontaneously. When h = 10b, for example, $\tau_{\max}$ reaches a value of G/40, which approaches the theoretical limit of shear strength in ideal crystals.

To clarify the mechanism of the "spontaneous increase in density" of the row one must investigate the fine structure of the field of stresses, τ_{xy} on both sides of the row. The boundaries of regions with different signs of τ_{xy} are shown schematically in Fig. 5d. In hatched areas tangential stresses continuously draw new dislocations to the row, and these locate themselves in the intervals between the dislocations already there.

One can make strict calculations of the field of stresses (25) of different dislocations. For an infinite row of equidistant dislocations one obtains (Cottrell [51])

$$\tau_{xy} = \frac{Gb}{(1-\nu)\,h}\;\frac{\dfrac{\pi x}{h}\left(\mathrm{ch}\,\dfrac{2\pi x}{h}\cdot\cos\dfrac{2\pi y}{h} - 1\right)}{\left(\mathrm{ch}\,\dfrac{2\pi x}{h} - \cos\dfrac{2\pi y}{h}\right)^2}. \tag{28}$$

Here the origin of the coordinates coincides with one of the dislocations; the X axis is directed along the Burgers vectors and the Y axis along the row.

In the plane passing through one of the dislocations of the row (y ± nh), we have

$$\tau_{xy} = \frac{Gb}{2\,(1-\nu)\,h}\,\frac{\dfrac{\pi x}{h}}{\text{sh}^2\,\dfrac{\pi x}{h}} \cdot \tag{29}$$

In the plane located between adjacent dislocations and passing exactly between them (y = h/2 ± nh), we have

$$\tau_{xy} = -\frac{Gb}{2\,(1-\nu)\,h}\,\frac{\dfrac{\pi x}{h}}{\text{ch}^2\,\dfrac{\pi x}{h}} \cdot \tag{30}$$

In the first case a dislocation of the same sign as the dislocations in the row is repelled, and in the second case it is attracted; the binding energy (the depth of the potential hole) is, regardless of the density of the row,

$$W = b\int_0^\infty \tau_{xy}\,dx = \frac{Gb^2}{2\pi\,(1-\nu)}\ln 2 \tag{31}$$

(calculated per unit length of the dislocation line). When b = 4 A, G = 4 · 10^{11} dynes/cm^2, and $\nu = \frac{1}{3}$ the binding energy turns out to be equal to approximately 10^{-4} erg/cm. These two particular solutions of Eq. (28) were previously investigated by Burger [55].

Thus under favorable conditions the interaction of edge dislocations of the same sign in parallel slip planes leads to the occurrence of vertical rows of edge dislocations. Burger [55] and Bragg [56] showed for the first time that the vertical row of dislocations located at a distance h from each other forms a boundary of blocks rotated symmetrically with respect to each other around an axis parallel to the dislocation by an angle

$$\theta = \frac{b}{h} \cdot \tag{32}$$

Expression (32) can be obtained by direct summation of lattice rotations (23) induced by single dislocations.

The processes described are particularly well-defined in the case of plastic bending of crystals. Wei and Beck [57] demonstrated in the case of a zinc crystal that vertical rows of dislocations (and of the corresponding rotations of the lattice around an axis in the slip plane and perpendicular to the direction of slip) occur directly during the process of plastic deformation and do not occur during later relaxation at room temperature. On the basis of Eq. (26) Wei and Beck showed that for each crystal there is a lower limit angle of disorientation of blocks. Since the acting stress must exceed the critical shear stress τ_{cr} of a given crystal, it follows from Eq. (28) that the distance between dislocations forming a vertical row must not exceed

$$h_{max} = \frac{Gb}{8\pi (1 - \nu) \tau_{cr}} ,\qquad (33)$$

which corresponds to the minimum possible angle of disorientation

$$\theta_{min} = \frac{b}{h_{max}} = \frac{\tau_{cr}}{G} 8\pi (1 - \nu). \qquad (33')$$

For zinc, for example, θ_{min} is approximately 0.5'.

However, during plastic bending of the crystal dislocations usually remain grouped in slip lines (Fig. 6a). It is only after later heat treatment,

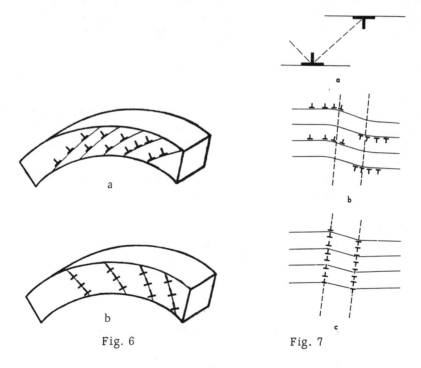

Fig. 6   Fig. 7

Fig. 6. Polygonization as a passage of dislocations from the horizontal rows (a) into vertical rows (b).

Fig. 7. Scheme of the formation of deformation bands. a) Diagonal equilibrium position of edge dislocations of different signs; b) mutual impedance of groups of dislocations of different signs; c) formation of edges of the deformation band as the result of polygonization.

which ensures not only the possibility of slipping of dislocations but also of
motion by diffusion perpendicular to the slip planes, that the horizontal rows
of dislocations in the slip planes can be destroyed and the vertical rows of
dislocations are formed (polygonization, Fig. 6b). By using the method of
selective etching, Cahn [58] and Hibbard and Dunn [59] obtained beautiful
illustrations of the correctness of the polygonization scheme shown in Fig. 6.

A detailed review of the relevant experimental investigations will be
found in [60] and [16].

INTERACTION BETWEEN EDGE DISLOCATIONS OF DIFFERENT SIGNS IN PARALLEL SLIP PLANES. FORMATION OF KINKS AND DEFORMATION BANDS

Equation (25) shows that tangential stresses τ_{xy} induced by an edge dislocation are equal to zero for three azimuths: $\vartheta = \pi/2$, $\vartheta = \pm \pi/4$. The
first case corresponds to the line located between the small petals of the
rosette of different tangential stresses in Fig. 5a and corresponds to the vertical equilibrium position of the edge dislocations of the same sign. The
azimuths $\vartheta = \pm \pi/4$ correspond to the lines located between the small and
large petals of the rosette of equal tangential stresses and corresponds to the
"diagonal" equilibrium position of edge dislocations of different signs located
in parallel slip planes (Fig. 7a).

The external tangential stress necessary to allow edge dislocations of different signs to pass each other in parallel slip planes without stopping in the
diagonal equilibrium position is given by formula (26). If the external stress
does not exceed τ_m the dislocation will not only stop but will stop all moving dislocations of the same sign which follow it. The accumulations of dislocations formed in this way will impede the slipping of dislocations in neighboring planes and favor the formation of new accumulations (Fig. 7b). It
must be noted that accumulations of positive dislocations will string themselves in one vertical row while accumulations of negative dislocations will
form another vertical row parallel to the first. As a result a deformation band
will gradually be formed with the positive dislocations on one side and the
negative dislocations on the other (Mott [61], Cottrell [42, 51]). After annealing, polygonization occurs, during which the accumulations of dislocations are dispersed and the deformation band becomes bordered by rows of dislocations alternately positive and negative (Fig. 7c). If the dislocation density is the same in both cases then the orientations of the lattice on both sides
of the deformation band will coincide and within the band the lattice will be
rotated at an angle determined by relationship (32).

The formation of kinks [62, 63] and other characteristic phenomena of
the reorientation of the lattice can be discussed on the basis of a mechanism
analogous to that of the formation of deformation bands. The relevant ex-

perimental results are described in detail in [60] and [16], and also by
Urusovskaya [64] in this collection. The main difference between a deformation band and a kink band is that in the first case the plastic deformation (motion of dislocations) occurs outside the band, and the lattice within the band
retains its original orientation. In the second case plastic deformation occurs
within the band, and the orientation of the lattice in the material surrounding
the band remains unchanged.

The quantitative theory of kink formation was developed by Frank and
Stroh [54]. The main results are based on the effects of the stress fields of
single edge dislocations. These authors represent the nucleus of the kink band
as a thin elliptical cylinder. Quantitative calculations show that for stresses
of the order of 10^{-3} G such a nucleus can grow if its length is approximately
10^{-3} cm and its thickness 10^{-5} cm. The concentration of stresses at the ends
of the band is then sufficient for the spontaneous formation of dislocations.
When the kink band reaches the boundary of the crystal it begins to grow rapidly in width. A number of ideas developed by the authors are also applicable to the process of growth of deformation bands. By use of a high speed
movie camera [108] it was shown that in spite of the assertions of Frank and
Stroh [54] the kinks occur not in regions with high shear stresses acting on the
slip planes in the direction of slip but, on the contrary, in regions where these
stresses decrease to zero and change sign. As a result, slip in the kink band
occurs in the direction opposite the direction of slip during the preceding
stage of plastic deformation.

Following Orowan [62] and Nye [36], Frank and Stroh examined the results obtained by Brilliantov and Obreimov [19], who discovered the rotation
of the lattice between double refraction bands in rock salt crystals as the result of the presence of deformation bands located perpendicular to the active
slip planes. Since the possible slip planes in NaCl crystals are perpendicular
to each other, the direction of slip bands and deformation bands can coincide,
and as a result become entangled. Apparently in the experiments of Stepanov,
Mil'kamanovich, Melankholin, and Regel' [37] the bands of double refraction
ascribed by these authors to slip bands are often accompanied by deformation
bands. From a theoretical and experimental investigation [109] it was possible to explain the mechanism of the formation of the bands found by
Brilliantov and Obreimov by a selection of active slip elements in adjacent
areas of the crystal. The stresses along the boundaries of the bands occur as
a result of incompatible plastic deformations in neighboring bands.

GENERAL CASE OF BLOCK FORMATION AS A RESULT OF PLASTIC DEFOR-
MATION

The particular cases of lattice reorientation described here were connected with the assembly of edge dislocations into vertical rows, which form the

so-called tilt boundaries of blocks, i.e., boundaries parallel to the axis of mutual rotation of blocks. Since, according to Eq. (23), each edge dislocation induces a rotation of the lattice around the direction of the dislocation line, block boundaries built of edge dislocations will always be tilt boundaries.

Screw dislocations have a very different effect. Relationship (24) shows that next to each screw dislocation the lattice is rotated around radial directions. Consequently a row of screw dislocations will create a rotation of the lattice around an axis perpendicular to the plane of the row forming the so-called twist boundary. To demonstrate this we can sum Eq. (24) for a row of screw dislocations or apply formula (2) directly to the change of the derivative of the displacement on both sides of such a row. However, a row of parallel screw dislocations induces not only a general rotation of the lattice but also macroscopic stresses [see formula (5)]. Therefore a stable twist boundary having the minimum possible energy for a given disorientation angle must be formed of at least two rows of screw dislocations (Fig. 8) selected in such a way that the rotations of the lattice induced by single rows add together while macroscopic stresses cancel each other (Frank [65], Van der Merwe and Frank [66]).

Frank [67] has developed a general theory which makes it possible to find the dislocation structure of a boundary when the orientation of the block boundaries and the axis of their mutual rotation are arbitrary. In particular, any boundary can be constructed (but in only one way) with three dislocation systems with noncoplanar Burgers vectors. For this case it is possible to calculate the direction and density of dislocation lines of each system (see Read [28]).

The initial equation of the theory for the case of a small disorientation of blocks has the form

$$S = \omega \times V, \tag{34}$$

where S is the sum of Burgers vectors of dislocations located at the boundary of blocks and intersected by the vector V in the plane of the boundary, and ω is the axial vector describing the mutual rotation of blocks (see [28], Fig. 12-4). Since S is a vector product, it is always perpendicular to V and ω for any values of V. Assuming that the boundaries of blocks occurring as the result of plastic deformation must be formed by dislocations located in their own slip planes, Ball and Hirsh [68] investigated the corresponding particular cases of Eq. (34).

If the blocks are formed as the result of plastic deformation along one slip system, all the dislocations must have a Burgers vector parallel to the direction of slip. The vector S will also be parallel to this direction. But since this vector is at the same time perpendicular to any vector V in the plane of the boundary, the boundary must be perpendicular to the direction of slip and, consequently, to the slip plane. The dislocation lines are located along the intersection between these planes and have an edge orientation. If we choose **V** parallel to the dislocation line we obtain **S** equal to zero and it then follows that the axis of the rotation of blocks is parallel to the dislocation lines. In other words, the boundary under investigation represents the previously mentioned vertical row of edge dislocations.

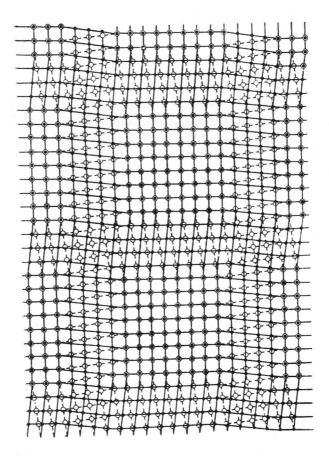

Fig. 8. Dislocational scheme of a tilt boundary. The small black and white circles represent atoms located above and below the boundaries of blocks, respectively. The axis of rotation of blocks is perpendicular to the plane of the figure. Screw dislocations form the quadrangular net.

Slip along two or more systems having a common slip plane is analyzed in an analogous way. If the boundary is not parallel to the slip plane it must be perpendicular to this plane. Then the axis of rotation is again parallel to the dislocation lines and the orientation of the boundary depends on the relationship between the density of dislocations with different Burgers vectors. If the boundary is parallel to the slip plane then V and S are always located in the same plane (the Burgers vector is always parallel to the slip plane). Therefore ω is always perpendicular to the plane of the boundary and the boundary is then not a tilt boundary but a twist boundary.

Slip along two or more systems with the same directions of slip corresponds to slip along a fixed direction, S (the sum of Burgers vectors). The boundary must be perpendicular to the direction of slip and must contain the axis of the rotation of blocks, i.e., it must be a tilt boundary. The orientation of the axis of rotation depends on the relationship between the densities of dislocations in different slip planes.

When slip occurs along two independent systems and the dislocations of the different systems are not parallel to each other, it is convenient to choose V_1 along the dislocations of one group. Then the corresponding vector S_1 will be parallel to the Burgers vector of the dislocations of the second group b_2 and consequently the axis of rotation must be perpendicular to vector b_2. In an analogous way it can be shown that the axis of rotation is perpendicular to the Burgers vector of the dislocations of the first group b_1, and consequently coincides with the direction of the normal to the plane of Burgers vectors. The direction of the dislocation line can also be determined uniquely: since the vector V_1 is perpendicular to S_1, the dislocation lines of the first group must be perpendicular to the Burgers vector of the dislocations of the second group, etc.

If in the case of slip along two independent systems all the dislocations are parallel to each other they must be parallel to the intersection line between the slip planes. If one chooses V along this line then $S = 0$. Consequently the axis of rotation of blocks is also parallel to the dislocation lines (the tilt boundary). But if the axis of rotation is in the plane of the boundary the vector S must always be perpendicular to the boundary. It follows then that the normal to the block boundary is parallel to the average Burgers vector of all the dislocations.

When slip occurs along different planes with coplanar directions of slip, S is always in the common plane of Burgers vectors and consequently the axis of rotation is perpendicular to this plane.

The determination of possible orientations of block boundaries and axes of rotation can be generalized to the case of three slip systems (Ball [69]). In this case the axis of rotation must be on the surface of some cone. A unique orientation of the block boundaries corresponds to each axis.

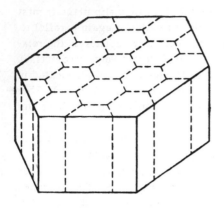

Fig. 9. Shape of the block character istic of hexagonal crystals. The base of the prism is formed by the twist boundary while the sides are formed by the rotation of boundaries. The dislocations are indicated by dashed lines.

We have shown in Fig. 9 the simplest type of a block theoretic- ally possible in hexagonal metals. Three slip systems have a common plane (basal). Typical blocks have the shapes of hexagonal prisms and are composed of hexagonal prisms; the bases of the prisms are parallel to the base of the block and are formed by twist boundaries com- posed of nets of screw dislocations. The side faces of the prisms are perpendicular to the direction of slip and represent tilt boundaries formed by the vertical rows of dis- locations with Burgers vectors par- allel to the corresponding direction of slip.

Ball and Hirsh [68, 69] and the reviews in [60] and [16] give a detailed comparison of the the- oretical and experimental results.

In a number of cases the simplified theory presented here must be made considerably more precise. For example, it is necessary to take into account possible deviations of the directions of dislocation lines from the slip planes as the result of the formation of steps on the dislocation lines (Read [28], Seitz [70]). In the addendum to the in- vestigation in [109] are considered incompatible boundaries (including stresses) of disoriented regions formed by dislocations located in their own slip planes. These stresses and the disorientation are described by the "coordinates" of the boundary in the space of the slip systems. Hyperplanes in the space of the slip planes which divide the regions corresponding to boundaries of different types are described.

DESCRIPTION OF THE GEOMETRY OF DEFORMED CRYSTALS

Nye [71] has investigated the geometry of a crystal subjected to hetero- geneous deformation. He proposed a direct method of describing the distor- tion of the crystal lattice based on dislocation concepts. The simplest result are obtained for plane deformation in a single slip system, which involves only one system of edge dislocations. In this case the slip planes are trans- formed into cylindrical surfaces and the distance between slip planes meas- ured along trajectories orthogonal to them remains unchanged (the author neglects macroscopic stresses). But then the curves corresponding to the traces of cylindrical slip surfaces have a common evolute (geometric locus of centers of curvature). Knowing this evolute one can construct a whole system of slip surfaces as a family of involutes. The tangents to the evolute are common normals to the slip surfaces (Fig. 10a). Consequently the curves orthogonal to the slip surface remain straight during deformation. They retain the same crystallographic direction as before deformation but do not remain parallel.

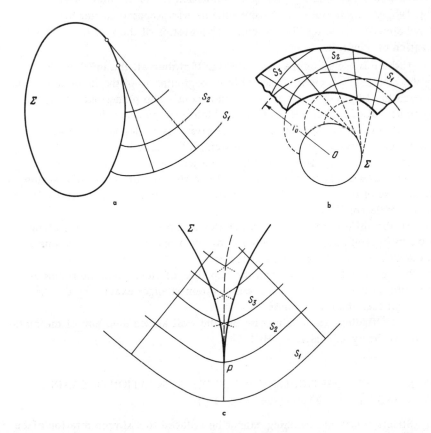

Fig. 10. Geometry of a deformed crystal according to Nye. a) Slip surfaces S_1, S_2, etc. are the involutes of the supporting curve, Σ; the normal to the slip plane remains straight; b) particular case of a circular bend—the slip planes are located along the involutes of the circle; c) the cusp, P on the supporting curve corresponds to the occurrence of a kink, and the surface of the kink (dashed line) terminates at point P.

If during polygonization the dislocations form rows perpendicular to the slip surfaces these rows will be straight (flat) not only in the microscopic but also in the macroscopic scale. For instance, if a hexagonal crystal is bent in a plane containing the axis c and one of the directions of basal slip, the slip plane must bend into cylinders with parallel generatrices while the boundaries of the polygons will be straight. In particular, in the case of a regular curved plane (Fig. 10b) the evolute is a circle. The slip surfaces are located along the involute of this circle. It was shown that in a sapphire (Kronberg [72]) and zinc (Gilman [73], Bilby and Smith [74]) the geometric of a plastically bent crystalline bar after polygonization is exactly that shown in Fig. 10b. A qualitatively analogous scheme was proposed earlier by Konobejewski and Mirer [75] in their interpretation of the results of polygonization of rock salt.

If there is a cusp on the evolute (Fig. 10c) then all the involutes formed by the converging regions of the evolute, beginning from the involute passing through the cusp, cross each other. Consequently slip planes undergo not a continuous bending but a sharp break, which corresponds to the surface of the kink. On the whole the geometry of the crystal is analogous to the well-known representation of the kink (Orowan [62], Urusovskaya [63, 64]).

In a similar way, Nye [71] derived the geometry resulting from more complex cases of plastic deformation. In particular, for plane deformation in the case of two orthogonal slip systems the geometry of the crystal follows three simple conditions:

a) the difference between the angles of rotation of the lattice along two slip lines of one system remains constant if one compares two points on the same slip line of the second system;

b) between two points along the slip lines of one system the radius of curvature of the slip line of the second system changes exactly by the distance between these two points.

These findings are nothing else but the well known theorems of the mathematical theory of plasticity (Hill [76]).

TWINNING AS A PARTICULAR CASE OF DISORIENTATION OF GRAINS. TWINNING DISLOCATIONS (see also [110])

Strictly speaking, twinning cannot be reduced to a simple rotation of the lattice, since of necessity it is accompanied by a mirror image of the lattice in the plane parallel to the twinning plane (twinning of the first order), or in the plane perpendicular to the direction of twinning (twinning of the second order). But in a crystal with a plane of symmetry the mirror image with respect to any plane is equivalent to the rotation of the lattice around the axis parallel to the intersection line between this plane and the plane of symme-

try of the lattice, and the twin boundary can be considered as a particular
case of the boundary of disoriented blocks.*

From the standpoint of energy, the twin in this case has the most advantageous orientation of the separation boundaries and the axis of rotation of
blocks (Read [28]).

If the twin boundary is a particular case of the tilt boundary of the grains
it can be arbitrarily represented as a series of edge dislocations (Bullough [77]).
As in the case of the slight disorientation of blocks described above, the axis
of rotation of the lattice is parallel to the dislocation lines. The twinning
plane cuts in two the angle between the slip planes of the components of the
twin, and the direction of twinning coincides with the bisectors of the obtuse
angle between the directions of slip (Fig. 11a).

The displacement of dislocations in their own slip planes leads to slip
parallel to the slip plane T, and the direction of slip (Fig. 11b). The magni-

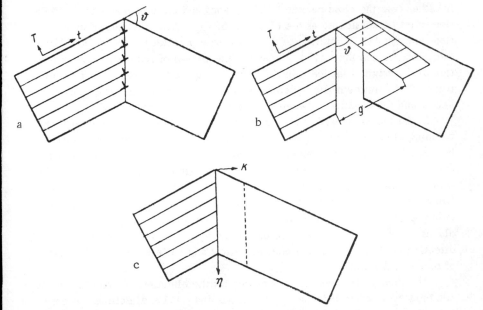

Fig. 11. Twinning as a particular case of rotation of blocks. a) Arbitrary
representation of the twin boundary as a row of edge dislocations located
in the slip plane of the original crystal; b) displacement of dislocation
induces plastic slip g; c) superimposition of the slip and the rotation of
the lattice is equivalent to twin slip parallel to the plane and the direction
of twinning.

---

* Of course the mechanism of formation of twins can be quite different
from the mechanism of the formation of disoriented blocks.

tude of slip g is related to the angle of rotation of the lattice ϑ by a simple relationship:

$$g = 2 \, \text{tg} \, \frac{\vartheta}{2} . \tag{35}$$

The slip g and the angle of rotation ϑ can finally be represented as slip parallel to the twinning plane K and the direction of twinning η (Fig. 11c).

In the case of first order twinning the lattice deformed by twin slip and then reflected in a twinning plane must coincide with the lattice of the original crystal. This is equivalent to the slip g and to the reflection in the plane perpendicular to the direction of slip t. In the case of second order twinning the lattice must be restored after the twin slip and the reflection in the plane perpendicular to the direction of twinning η, which is equivalent to slip g and the reflection in the slip plane T.

Bullough applied this theory to face-centered cubic and diamond lattices. In either case the close packed $\{111\}$ planes and close packed $<110>$ directions must be considered as the elements of slip. Also, the line of dislocations and the axis of rotation of the lattice are parallel to one of the $<112>$ directions. Since the $\{110\}$ planes are the planes of symmetry, the restoration of the original lattice resulting from slip g must correspond to first order twins. The minimum slip is determined by the ratio between the translation vector and the distance between the closest $\{111\}$ planes.

In the face-centered cubic lattice $g_{min} = \sqrt{6}/2$, which corresponds to twinning along the $\{113\}$ plane. In the diamond lattice the closest distance between $\{111\}$ planes is one-fourth that in the face-centered cubic lattice and, correspondingly, the minimum slip is four times larger, i.e., $g_{min} = 2\sqrt{6}$, which corresponds to twinning along the $\{123\}$ plane. This result was confirmed in the experiment. A slip twice as large as g_{min} corresponded to twinning along the $\{345\}$ plane, etc. The deformation twins along the $\{111\}$ planes in the diamond lattice are explained by taking into account the $<112>$ direction of slip. This model does not allow any variation for the formation of second order slip.

The theory also gives correct results for the elements of twinning in body-centered cubic crystals if the $\{112\}$ planes and $<111>$ directions are considered as elements of slip. The arbitrary representation of the twin boundary as a row of slipping dislocations was used by Bullough and Bilby [78] in their investigation of martensite transformations, which are accompanied by complex lattice deformation.

Let us emphasize once again that in the examples cited the slip boundary is arbitrarily represented as a row of ordinary dislocations since in these cases the angle of rotation of the lattice corresponds to the distance between dislocations of the order of or less than the width of the dislocation, and therefore the concept of separate dislocation lines loses its meaning. Therefore we cannot assume that the motion of the twin boundary occurs as a result of slipping of dislocations according to the scheme given in Fig. 11.

The dislocation theory of the mechanism of displacement of these twin boundaries can be developed independently of the concept developed previously that twins can be reduced to the particular case of disorientation of blocks.

The dislocation theory, which is in complete accord with the concept developed by Frenkel' and Kontorova [79], Vladimirskii [80], and Lifshits and Obreimov [81], considers that the growth of one of the components of the twin at the expense of the other occurs by translational displacement of atomic steps on the surface of separation of twins. Such steps correspond to the edge of an incomplete plane of atoms which have passed into the twinning position, and are called twinning or partial dislocations. (For a more detailed discussion of dislocational theories of twinning see [100]).

The scheme of the polar mechanism of the growth of twins proposed by Cottrell and Bilby [83] was discussed in detail in [82]. In body-centered cubic crystals the twinning dislocation rotating around an ordinary dislocation crossing the twin boundary increases the thickness of one of the components of the twin by one layer at each revolution (if the thickness of the layer corresponds to the screw component of the Burgers vector of an ordinary dislocation). Lately Ookawa [84] proposed a similar but more complex scheme of a polar mechanism of twin growth in face-centered cubic crystals. In his opinion such a mechanism could explain the experimentally observed low value of critical stress sufficient for twinning of copper at helium temperatures.

Analysis of the relationship between dislocations and twins makes it possible to assume that ordinary incomplete dislocations located in close-packed planes can split into partial twinning dislocations, with the formation of intermediate layers composed of one-layer twins—stacking faults (Read [28], Seeger and Schoeck [85], Lomer [86]). Recent investigations by Hirsch, Horne, Whelan, and Bollmann [7, 8] confirmed this phenomenon predicted by the theory. Recently stacking faults and partial dislocations of different types, including not only half dislocations but quarter dislocations, have been found in a whole series of materials.

3. DISLOCATIONAL DESCRIPTION OF MACROSCOPIC DE-FORMATIONS

To account for dislocations in the macroscopic theory of plasticity it is necessary to describe the distribution of dislocations in macroscopic terms. Let us consider the average distribution of dislocations over an area crossed by a great number of single dislocation lines. Indenbom and Orlov [107] give a different and more logical treatment of the macroscopic dislocation theory by using distribution functions, $\mathbf{f_b}$ $(\mathbf{z}, \mathbf{n})$, of the dislocations with a Burgers vector $\mathbf{b}$ along the coordinate $\mathbf{z}$ and direction $\mathbf{n}$. Then the total density of dislocations (the total length of dislocation lines per unit volume) is $N =$ $= \int (d\mathbf{b})(d\mathbf{n})f_b$, the tensor of macroscopic density of dislocations is $\beta_{ij}=\int(d\mathbf{b}) \cdot$ $\cdot \int(d\mathbf{n})b_i n_j f_b$, and the tensor of motion of dislocations is $\int(d\mathbf{b})(d\mathbf{n}) = v_i b_j n_k f_b$, where v is the velocity of dislocations, etc. Nye [71], Bilby [87], and Kröner [88] showed that macroscopic dislocation density in this case is characterized by a tensor of the second rank β, whose ijth component is equal to the jth component of the sum of Burgers vectors of all dislocations which cross a unit area perpendicular to axis i.

The well-known geometric rule of finding the Burgers vector by the distance between the ends of the contour surrounding the dislocations is equivalent to finding the contour integral of elastic displacements. In an analogous way, the tensor of macroscopic dislocation density corresponds to the circulation of the elastic distortion ε composed of the symmetric tensor of elastic deformation e and of the antisymmetric tensor of the rotation of the lattice, ω, equivalent to the axial vector of rotations $\boldsymbol{\omega}$

$$\beta = - \operatorname{Rot} \varepsilon, \tag{36}$$

$$\varepsilon = e + \omega = e - \omega \times I \tag{37}$$

(I is a unit tensor).

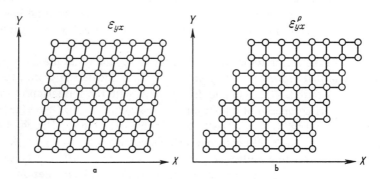

Fig. 12. Variation of the shape of the crystal as the result of purely elastic (to the left) and plastic (to the right) slip distortion. The lattice sites are represented by circles.

The diagonal terms ε correspond to elastic elongation (compression), the nondiagonal to elastic slip, the first index to the plane, and the second index to the direction of slip. An example of slip ε_{yx} is given in Fig. 12a.

In the case of a simple cubic lattice the diagonal terms of the tensor β describe the density of screw dislocations and the nondiagonal describe the

density of edge dislocations. In more complex cases such treatment is not
applicable and it is necessary to isolate within the tensor of macroscopic dislocation density the contribution of dislocations with all Burgers vectors possible in this particular lattice (Indenbom [89]).

PLASTIC DEFORMATION AND DISLOCATION DENSITY

We have derived above a simple relationship between the plastic change
of shape and the motion of dislocations in one isolated slip plane [Eq. (1)].

In the general case the motion of dislocations can be described by a
third rank tensor N_{ijk}, where i is the direction of the motion of dislocations,
j the direction of dislocation lines, and k the direction of Burgers vector
(Kröner [90]). In a simple cubic lattice $k = i \neq j$ corresponds to the slip of
edge dislocations, $i \neq j = k$ to slip of screw dislocations, and $i \neq j \neq k \neq i$ to
the motion of edge dislocations.

The slipping of dislocations induces plastic slip parallel to the slip plane
in the direction of the Burgers vector. The motion of edge dislocations is
equivalent to the shortening (or elongation) of corresponding excess planes
and leads to plastic shortening or elongation of the crystal in the direction
parallel to the Burgers vector. In the general case the macroscopic plastic
distortion ε^P can be related to the tensor of the displacement of dislocations
in the following way (Kröner [90]):

$$\varepsilon_{ij}^p = b\epsilon_{mni}N_{mnj} \tag{38}$$

or

$$\epsilon_{ijl}\varepsilon_{lk} = b\,(N_{ijk} - N_{jik}). \tag{38'}$$

Here the ϵ_{ijk} tensor is an antisymmetric tensor of the third rank ϵ_{ijk} is equal
to +1 if i, j, and k represent the even commutations of indices 1, 2, and 3.
The repetition of indices is considered to be summation.

The plastic distortion can be divided into a symmetrical part— tensor of
plastic deformation ε^P— and an antisymmetrical part— tensor of plastic deformation ω^P, which is equivalent to the vector of rotation $\boldsymbol{\omega}^P$

$$\omega_i^p = \frac{b}{2}\,(N_{jij} - N_{ijj}). \tag{39}$$

Figure 12b illustrates the plastic distortion ε_{yx}^P. The first index corresponds
to the plane, and the second to the direction of slip. Comparison of Figs. 12a
and 12b shows the difference between purely elastic and purely plastic distortion. In the first case the coordination of atoms is preserved but the bonds
are under tension, and as a result elastic deformation and rotation of the lattice occur. In the second case the atoms are displaced into new equilibrium
positions, the bonds are not under tension, and there is no rotation of the lat-

tice due to the translational character of the displacements; there is only a change in the external shape of the body, which can be described in terms of plastic elongation and plastic change of angles.

If the displacement of dislocations is irregular, dislocations with a density

$$\beta_{jk} = -b \frac{\partial}{\partial x_i} N_{ijk} \tag{40}$$

will remain in the crystal.

By differentiating Eq. (38') along the j coordinate we see that the density of dislocations is determined by the circulation of plastic distortion

$$\beta = \text{Rot } \varepsilon^p. \tag{41}$$

Comparison of Eqs. (36) and (41) shows that the total distortion $\varepsilon + \varepsilon^P$ is irrotational:

$$\text{Rot } (\varepsilon + \varepsilon^p) = 0. \tag{42}$$

Consequently, the total distortion can be represented in terms of a gradient of some vector of total displacement $\mathbf{u}$:

$$\varepsilon + \varepsilon^p = \text{Grad } \boldsymbol{u}. \tag{43}$$

If there are no dislocations then the vectors of elastic displacement [usual practice in the theory of plasticity and of plastic slip can be found by Eqs. (36) and (41), respectively]. In the general case such vectors do not exist and only the vector of the total displacement can be determined.

DISLOCATIONS AND THE CURVATURE OF THE LATTICE

Let us again consider Eq. (36) and separate the symmetric and antisymmetric parts in the expression of elastic distortion. Using Eq. (37), we have

$$\beta = -\text{Rot } e + \frac{d\boldsymbol{\omega}}{dx} - I \text{ div } \boldsymbol{\omega}. \tag{44}$$

It follows that the sum of diagonal elements β, which represents the total density of screw dislocations, is determined by the divergence of the vector of rotation, which in the presence of dislocations is no longer equal to zero:

$$\text{Sp } \beta = -2 \text{ div } \boldsymbol{\omega}. \tag{45}$$

Tensor $\varkappa = d\boldsymbol{\omega}/dx$, corresponding to tensor Grad $\boldsymbol{\omega}$, describes the curvature of the lattice. If the lattice curvature and the macroscopic elastic deformation of the crystal are found experimentally (e.g., by x-ray or optical measurements), relationship (44) makes it possible to calculate the macroscopic dislocation density inducing a given field of distortions of the lattice.

If the macroscopic elastic deformations (and stresses) are absent, then a simple relationship between the dislocation density and the curvature of the lattice follows from (44) and (45), namely,

$$\left.\begin{aligned}
\beta &= \varkappa - I \operatorname{Sp} \varkappa, \\
\varkappa &= \beta - \frac{1}{2} I \operatorname{Sp} \beta.
\end{aligned}\right\} \tag{46}$$

These relationships were derived for the first time by Nye in the publication mentioned earlier [71]. In particular, it is easy to see that the deviators of the tensors of lattice curvature and of the density of dislocations are equal to each other. If the spherical parts of the tensor, β and $\varkappa$, are equal to zero the tensors themselves are equal, and we have

$$\beta = \varkappa. \tag{46'}$$

It is just this consequence which is usually used to check the theory (see[17]).

PLANE DEFORMATION

The general equations are considerably simplified in the case of plane deformation. Let us consider, for example, that all the displacements occur perpendicular to axis Z. Only two components will remain in the tensor of the density of dislocations (ω corresponds to ω_z):

$$\left.\begin{aligned}
\beta_{zx} &= \frac{\partial}{\partial y} e_{xx} - \frac{\partial}{\partial x} e_{xy} + \frac{\partial \omega}{\partial x}, \\
\beta_{zy} &= - \frac{\partial}{\partial x} e_{yy} + \frac{\partial}{\partial y} e_{xy} + \frac{\partial \omega}{\partial y}.
\end{aligned}\right\} \tag{47}$$

By using Hooke's Law and the equation of equilibrium stress one can pass from the deformations to stresses by excluding all the components of the stress tensor except the sum of the main stresses

$$\left.\begin{aligned}
\beta_{zx} &= \frac{1 - \nu^2}{E} \frac{\partial}{\partial y} (\sigma_x + \sigma_y) + \frac{\partial \omega}{\partial x}, \\
\beta_{zy} &= - \frac{1 - \nu^2}{E} \frac{\partial}{\partial x} (\sigma_x + \sigma_y) + \frac{\partial \omega}{\partial y}.
\end{aligned}\right\} \tag{48}$$

In regions with a constant pressure the dislocation density is determined by the lattice curvature. In regions of the crystal where there are no dislocations (more precisely, the macroscopic density of dislocations is equal to zero) the curvature of the lattice determines the pressure gradient.

In the case of a cylindrical bend of an infinite plate whose surfaces are perpendicular to the y axis, $\sigma_y = 0$, and all the values depend only on the y coordinate.

As a result the fundamental equation developed by Read on the basis of the dislocation theory of plastic bending [91] follows immediately from relationship (48) (it is assumed that $\varkappa = \partial \omega / \partial x$):

$$\beta_{zx} = \frac{1-\nu^2}{E}\frac{\partial}{\partial y}\sigma_x + \varkappa. \tag{49}$$

In the elastic region next to the neutral axis $\beta = 0$ and $\sigma_x = y\varkappa E/(1-\nu^2)$. In plastic zones $\sigma_x = \text{const.}$, while dislocation density corresponds to curvature of the crystal, $\beta_{zx} = \varkappa$.

DISLOCATIONAL TREATMENT OF INCOMPATIBILITY OF DEFORMATIONS

According to the theory of elasticity it is possible to calculate the displacement (with a precision of the displacement and rotation of the body as a whole) by the known deformations if the known deformations satisfy the Saint Venant conditions of compatibility. If in some region O (see Fig. 13a)

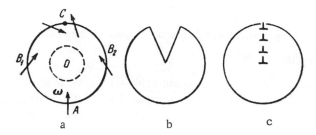

a   b   c

Fig. 13. Relationship between dislocation and the incompatibility of deformation. a) Passage around the deformed region, O along AB_1C and AB_2C gives different rotation vectors for point C; b) cutting of a wedge and joining the edges of the cut creates incompatibility of deformation at the origin of the wedge; c) representation of a wedge dislocation in the form of a broken row of Burgers dislocations.

the conditions of compatibility are not satisfied, the vector of rotation ω, cannot be constructed uniquely from the deformation, and if region O is traced along two different ways, AB_1C and AB_2C, two different values of ω are obtained for the same point, C.

Figure 13b shows how the ramification of the vector of rotations can occur when the region of incompatibility is encircled. Let us cut a wedge from a disc and then join the edges of the cut. We thus obtain a Volterra dislocation corresponding to the rational boundaries of blocks which terminate in the incompatibility region (Fig. 13c). In other words, the incompatibility of deformations can be regarded as the result of a distribution of broken rotational boundaries of blocks within a body [92].

The formula which relates the incompatibility of deformations to the distribution of dislocations was first given by Kröner [88]. In the absence of dislocations the ordinary elasticity theory gives directly from Eq. (44)

$$\frac{d\boldsymbol{\omega}}{dx} = \text{Rot } e \tag{50}$$

(div ω = 0) or transformed:

$$(\text{Rot } e)^* = \text{Grad } \boldsymbol{\omega} \tag{50'}$$

(the asterisk indicating transformation).

In order for the differential equation (50') to make it possible to reeestablish uniquely the vector of rotation ω by the known deformation, the tensor (Rot **e**)* must be irrotational. Assuming that

$$\text{Rot } (\text{Rot } e)^* = -S, \tag{51}$$

we obtain Saint Venant conditions of compatibility in the form

$$S = 0, \tag{52}$$

where S is the so-called tensor of the incompatibility of deformations. Figure 13a, b, and c corresponds to $S_{zz} \neq 0$, if we consider axis Z perpendicular to the figure. The other components of S also correspond to Volterra dislocations and can be represented in the form of broken rows (or nets) of Burgers dislocations (screw and edge) [92].

When dislocations are present we can take into account relationship (44) and, using the identity

$$\text{Rot } \varphi - (\text{Rot } \varphi)^* = (\text{Div } \varphi^* - \text{grad Sp } \varphi) \times I, \tag{53}$$

we obtain from Eq. (51) [88]

$$S = \text{Rot } \left(\beta^* - \frac{1}{2} I \text{ Sp } \beta\right) = \frac{1}{2} [\text{Rot } \beta^* + (\text{Rot } \beta^*)^*]. \tag{54}$$

The incompatibility of deformations is determined by the irregularities in the distribution of dislocations, and, consequently, no internal stresses (macroscopic) can exist when the distribution of dislocations is regular.

Let us note that the dislocational concept of incompatibility of deformations, which completely determines the internal stress, is applicable not only to crystalline but also amorphous bodies. See the effect functions for incompatibility of deformations and dislocations in [111].

CONCLUSION

The examples of dislocational descriptions of separate characteristic phenomena of plastic deformation examined here and the interpretation of the general concept of the macroscopic theory of elasticity and plasticity in dislocational terms showed the wealth of information resulting from the attempt to use the dislocation theory in analyzing concrete processes of plastic deformation and the possibility of correlating the dislocation theory (as a microscopic theory) with the macroscopic theory of the mechanism of displacement of atoms.

On the other hand, it is impossible to derive a theory of the mechanical properties of crystals on the basis of the dislocation theory per se. The large variety of possible types of collective displacements of atoms predicted by the dislocation theory makes it difficult to solve the question of which mechanism is the most important in a given process. This question can be answered only by experiments and comparison of the data with those predicted by the dislocation theory.

Only by comparing qualitative and quantitative experimental and theoretical results can one proceed from checking the dislocation hypothesis to the construction of a theory of plasticity and strength, and determine precisely the limits of applicability of the dislocation concept. The present state of the dislocation theory makes it possible to undertake such an investigation.

Added note: In the four years since the present article was written the physics of strength and plasticity has made considerable progress, although the main progress has been in the theory of dislocation itself and not in the dislocation theory of plastic deformation and destruction. The present state of the problem and the current literature are discussed in [101] and [107].

LITERATURE CITED

1. J. W. Menter, "The direct study by electron microscopy of crystal lattices and their imperfections," Proc. Roy. Soc. A. 236, 119-315 (1956).
2. H. Hashimoto and R. Uyeda, "Detection of dislocation by the moiré pattern in electron micrographs," Acta cristallogr. 10, 143 (1957).
3. D. W. Pasley, J. W. Menter, and G. A. Basset, "Observation of dislocations in metals by means of moiré patterns on electron micrographs," Nature 179, No. 4563, 752-755 (1957).
4. A. V. Shubnikov, "Moiré," Priroda 16, 83-88 (1927).
5. A. V. Shubnikov, Crystals in Science and Technology [in Russian] (Izd. AN SSSR, 1956) p. 9.
6. P. B. Hirsh, R. W. Horne, and M. J. Whelan, "Direct observations of the arrangement and motion of dislocations in aluminum, Phil. Mag. 1, 677-684 (1956).
7. W. Bollmann, "Interference effects in the electron microscopy of thin crystal foils," Phys. Rev. 103, 1588-1589 (1956).
8. M. J. Whelan, P. B. Hirsh, R. W. Horne, and W. Bollmann, "Dislocations and stacking faults in stainless steel," Proc. Roy. Soc. A, 240, 523-538 (1957).

9. F. L. Vogel, W. G. Pfann, H. E. Corey, and E. E. Thomas, "Observations of dislocations in lineage boundaries in germanium," Phys. Rev. 90, 489-490 (1953).

10. F. L. Vogel, "Dislocations in low-angle boundaries in germanium," Acta metallurg. 3, 245-248 (1955); "Dislocations in plastically bent germanium crystals," Trans. AIME 206, 946-949 (1956).

11. S. Amelinckx, "Etch pits and dislocations along grain boundaries, slip lines and polygonization walls," Acta metallurg. 2, 843-853 (1954).

12. S. Amelinckx, "The direct observation of dislocation nets in rock salt crystals," Phil. Mag. 1, 269-290 (1956).

13. W. C. Dash, "Copper precipitation in silicon," J. Appl. Phys. 27, 1193-1195 (1956).

14. V. L. Indenbom and G. E. Tomilovskii, "Internal stresses around single dislocations," Doklady Akad. Nauk SSSR 115, 723-726 (1957).

15. A. J. Forty, "Direct observations of dislocations in crystals," Advances Phys. 3, 1-25 (1954).

16. P. B. Hirsh, "Mosaic structure," Progress in metal physics 6, 236-339 (1956).

17. V. L. Indenbom, "Dislocations in crystals," Kristallografiya 3, 113-132 (1958); S. Vonsovskii and A. Orlov, "Dislocations in crystals," Probl. Sovr. Fiz. No. 9 (1957).

18. I. V. Obreimov and L. V. Shubnikov, "Investigation of plastic deformation in rock salt crystals by optical methods," Zhur. Russ. fiz. khim. Obsh. 58, 817 (1926).

19. N. A. Brilliantov and I. W. Obreimov, "On plastic deformation of rock salt," Phys. Z. Sowjetunions 6, 587-602 (1934); 12, 7-19 (1937).

20. E. Orowan, "Zur Kristallplastizität," Z. Phys. 89, 605, 614, 634 (1934).

21. M. Polanyi, "Über eine Art Gitterstörung, die einen Kristallplastisch machen konnte," Z. Phys. 89, 660 (1934).

22. G. I. Taylor, "The mechanism of plastic deformation of crystals," Proc. Roy. Soc. A, 145, 362-387 (1934).

23. T. A. Kontorova and Ya. I. Frenkel', "The theory of plastic deformation in twinning," Zhur. éksp. teor. fiz. 8, 89-95, 1340-1348 (1938).

24. M. V. Klassen-Neklyudova and T. A. Kontorova, "Modern theoretical concepts of the nature of plastic deformation," Usp. fiz. nauk 26, 217-237 (1944); "The dislocation hypothesis of plasticity," Usp. fiz. nauk 52, 143-151 (1954).

25. A. V. Stepanov, "Dislocation theories of strength and plasticity," Izv. AN SSSR, OTN No. 9, p. 90-107 (1954).

26. R. I. Garber, I. V. Obreimov, and L. M. Polyakov, "Formation of ultramicroscopic heterogeneities during plastic deformation of rock salt," Doklady Akad. Nauk SSSR 108, 425-427 (1956).

27. A. A. Bochvar and Yu. A. Preobrazhenskaya, "The nature of slip lines visible under the microscope, "Doklady Akad. Nauk 113, 564-566 (1957).

28. W. T. Read, Dislocations in Crystals [Russian translation] (Metallurgizdat, 1957).

29. V. L. Indenbom, "Mobility of dislocations in the Frenkel'-Kontorova model," Kristallografiya 3, 197-205 (1958).

30. R. Peierls, "The size of a dislocation," Proc. Phys. Soc. A, 52, 34-37 (1940).

31. F. R. Nabarro, "Dislocations in simple cubic lattice," Proc. Phys. Soc. A, 59, 256-272 (1947).

32. G. Leibfried and K. Lücke, "Über das Spannungsfeld einer Versetzung," Z. Phys. 126, 450-464 (1949).

33. J. D. Eshelby, "Edge dislocations in anisotropic materials," Phil. Mag. 40, 903-912 (1949).

34. H. B. Huntington, "Modification of Peierls — Nabarro model for edge dislocation core," Proc. Phys. Soc. B, 68, 1043-1048 (1955).

35. H. B. Huntington, J. E. Dickey, and R. Thomson," Dislocation energies in NaCl," Phys. Rev. 100, 1117-1128 (1955).

36. J. F. Nye, "Plastic deformation of silver chloride. Photoelastic study of internal stresses in glide pockets," Proc. Roy. Soc. A, 200, 47-66 (1949).

37. A. V. Stepanov, "Phenomena of artificial slip formation," Zhur. éksp. teor. fiz. 17, 601-606 (1947); 18, 741-749 (1948); A. V. Stepanov and E. A. Mil'kamanovich, "Phenomena of artificial slip formation," Zhur. éksp. teor. fiz. 18, 773-775 (1948); 21, 401-408, 409-412 (1951); N. M. Melankholin and V. R. Regel', "An experimental investigation of slip formation in NaCl crystals," Trudy, In-ta Kristallogr. AN SSSR No. 12, 148-157 (1956).

38. V. E. Kochnov and M. P. Shaskol'skaya, "Investigation of slip lines in silver chloride crystals," Kristallografiya 2, 274-277 (1957).

39. V. L. Indenbom and G. E. Tomilovskii, "Macroscopic dislocations in a corundum crystal," Kristallografiya 2, 190-192 (1957).

40. O. Mügge, "Über Translationen und verwandte Erscheinungen in Kristallen. Biegegleitung," Neues Jahrb. Mineral. 1, 155-158 (1898).

41. M. Polanyi, "Biegegleitung," Z. Kristallogr. 61, 49 (1925).

42. A. H. Cottrell, "Theory of dislocations," Progr. Metal Phys. 1, 77 (1949).

43. J. D. Eshelby, F. C. Frank, and F. R. Nabarro," Equilibrium of linear arrays of dislocations," Phil. Mag. 42, 351 (1951).

44. G. Leibfried, "Verteilung von Versetzungen in statischem Gleichgewicht," Z. Phys. 130, 214-226 (1951).

45. P. A. Jacquet, "Recherches experimentales sur la microstructure de la solution solide cuivre-zinc 65/35," Acta metallurg. 2, 752-790 (1954).

46. B. A. Bilby and A. R. Entwisle, "Dislocation arrays and rows of etchpits," Acta metallurg. 4, 257-261 (1956).
47. A. H. A. Meleca, "Etchpits and dislocations in zinc single crystals," Phil. Mag. 1, 803-811 (1956).
48. J. J. Gilman, "Etch pits and dislocations in zinc monocrystals," Trans. AIME 206, 998-1004 (1956).
49. A. N. Stroh, "The formation of cracks as result of plastic flow." Proc. Roy. Soc. A, 223, 404-414 (1954); "Brittle fracture and yielding," Phil. Mag. 46, 968-972 (1955).
50. N. J. Petch, "The fracture of metals," Progr. Metal Phys. 5, 1-52 (1954); N. F. Mott, "The fracture of metals," J. Iron and Steel, Inst. 183, 233-243 (1956).
51. A. H. Cottrell, Dislocations and Plastic Flow in Crystals [Russian translation] (Metallurgizdat, 1958).
52. W. L. Bond and J. Andrus, "Photographs of the stress field around edge dislocations," Phys. Rev. 101, 1211 (1956).
53. V. L. Indenbom and M. A. Chernysheva, "Loop of an edge dislocation in crystals of tartaric acid," Doklady Akad. Nauk SSSR 111, 596-598 (1956).
54. F. C. Frank and A. N. Stroh, "On the theory of kinking," Proc. Phys. Soc. B, 65, 811-821 (1952).
55. J. M. Burgers, "Some considerations on the fields of stress connected with dislocations in a regular crystal lattice," Proc. Kon. Ned. Acad. Wetenschap. Amst. 42, 293-325, 378-399 (1939).
56. W. L. Bragg, Proc. Phys. Soc. 52, 54-55 (1940).
57. C. T. Wei and P. A. Beck, "Structure of bent zinc crystals," J. Appl. Phys. 27, 1508-1518 (1956).
58. R. W. Cahn. "Recrystallization of single crystals after plastic bending," Inst. Metals 76, 121 (1949-1950).
59. W. H. Hibbard and C. G. Dunn, "A study of <112> edge dislocation in bent silicon-iron single crystals," Acta metallurg. 4, 306-315 (1956).
60. R. Maddin and N. K. Chen, "Geometrical aspects of the plastic deformation of metal single crystals," Progr. Metal Phys. 5, 53-95 (1954).
61. N. F. Mott, "The mechanical properties of metals," Proc. Phys. Soc. B, 64, 729-741 (1951).
62. E. Orowan, "A new type of plastic deformation in metals," Nature 149, No. 3788, 643-644 (1942).
63. A. A. Urusovskaya, Study of Complex Phenomena of the Plastic Deformation of Crystals, Dissertation (In-t. Kristallografii AN SSSR, 1955).
64. A. A. Urusovskaya, "Formation of regions with a reoriented lattice resulting from deformation of mono- and polycrystals," Nastoyashchii Sbornik pp. 67-116.

65. F. C. Frank, "Bristol Conference on Strength of Solids, 1947 (London, 1948) p. 46.

66. J. H. Van der Merwe and F. C. Frank, "Misfitting monolayers," Proc. Phys. Soc. A, 62, 315-316 (1949).

67. F. C. Frank, "Plastic deformation of crystalline solids," Carnegie Inst. Technology and U. S. Office of Naval Res., Pittsburgh, 1950, pp. 150-151.

68. C. J. Ball and P. B. Hirsh, "Surface distributions of dislocations in metals," Phil. Mag. 46, 1343-1352 (1955).

69. C. J. Ball, "Surface distributions of dislocations in metals," Phil. Mag. 2, 977-984 (1957).

70. F. Seitz, "On the generation of vacancies by moving dislocations," Advances Phys. 1, 43-90 (1952).

71. J. F. Nye, "Some geometrical relations in dislocated crystals," Acta metallurg. 1, 153-162 (1953).

72. M. L. Kronberg, "Polygonization of a plastically bent sapphire crystal," Science 122, 599-600 (1955).

73. J. J. Gilman, "Structure and polygonization of bent zinc monocrystals," Acta metallurg. 3, 277-288 (1955).

74. B. A. Bilby and E. Smith, "The shapes of glide planes on bent zinc crystals," Acta metallurg. 4, 379-381 (1956).

75. S. Konobejewski and I. Mirer, "Die röntgenographische Bestimmung elastischer Spannungen in gebogenen Kristallen," Z. Kristallogr. 81, 69 (1932).

76. R. Hill, Mathematical Theory of Plasticity [Russian translation] (IL, 1950) p. 164.

77. R. Bullough, "Deformation twinning in the diamond structure," Proc. Roy. Soc. A, 241, 568-577 (1957).

78. R. Bullough and B. A. Bilby, "Continuous distribution of dislocations. Surface dislocations and the crystallography of martensitic transformations," Proc. Phys. Soc. B, 69, 1276-1286 (1956).

79. T. A. Kontorova and Ya. I. Frenkel', "The theory of plastic deformation and twinning," Zhur. éksp. teor. fiz. (1938) pp. 1349-1358.

80. K. V. Vladimirskii, "Twinning in calcite," Zhur. éksp. teor. fiz. 17, 530-536 (1947).

81. I. M. Lifshits and M. V. Obreimov, "Some considerations of twinning in calcite," Izv. AN SSSR, Ser. Fiz. 12, 65-80 (1948); I. M. Lifshits, "Macroscopic description of the twinning of crystals," Zhur. éksp. teor. fiz. 18, 1134-1143 (1948).

82. A. H. Cottrell, "Theory of dislocations," Progr. Metal. Phys. 4, 205 (1953).

83. A. H. Cottrell and B. A. Bilby, "A mechanism for the growth of deformation twins in crystals," Phil. Mag. 42, 573-581 (1951).

84. A. Ookawa, "On the mechanism of deformation twin in fcc crystal," J. Phys. Soc. Japan 12, 825 (1957).

85. A. Seeger and G. Schoeck, "Die Aufspaltung von Versetzungen in Metallen dichtester Kugelpackung," Acta metallurg. 1, 519-530(1953).

86. W. M. Lomer, "A dislocation reaction in the face-centered cubic lattice," Phil. Mag. 42, 1327-1331 (1951).

87. B. A. Bilby, "Types of dislocation sources," Reports Conference Defects in Crystalline Solids, Bristol, 1954 (London, 1955) pp. 124-133.

88. E. Kröner, "Der fundamentale Zusammenhang zwischen Versetzungsdichte und Spannungsfunktionen," Z. Phys. 142, 463-475 (1955).

89. V. L. Indenbom, "Macroscopic theory of the formation of dislocations during crystal growth," Kristallografiya 2, 594-603 (1957).

90. E. Kröner and G. Rieder, "Kontinuumstheorei der Veztsetzungen," Z. Phys. 145, 424-429 (1956).

91. W. T. Read, "Dislocation theory of plastic bending," Acta metallurg. 5, 83-88 (1957).

92. J. D. Eshelby, "The continuum theory of lattice defects," Solid State Phys. (1956) 3, 79-145.

93. "Dislocations in crystals," Probl. Sovr. Fiz. No. 9 (1957).

94. "Dislocations in crystals," Bibliograficheskii Ukazatel' (Izd. AN SSR, 1959). 1959).

95. Dislocations and Mechanical Properties of Crystals (Conf. Lace Placid, 1956) (N. Y., 1957).

96. Internal Stresses and Fatigue in Metals (Sympos., Detroit, 1958) (N. Y., 1959).

97. S. Amelinckx and W. Dekeyser, "The structure and properties of grain boundaries," Solid State Physics 8, 325-499 (1959).

98. P. B. Hirsh, "Observations of dislocations in metals by transmission electron microscopy," J. Inst. Metals 87, 406-419 (1959).

99. V. R. Regel', A. A. Urusovskaya, and V. N. Kolomiichuk, Kristallografiya 4, 937-955 (1959).

100. P. B. Hirsh, "Direct experimental evidence of dislocations," Metallurgical Rev. 4, 101-140 (1959).

101. M. V. Klassen-Neklyudova and V. L. Indenbom, Introduction to a collection of translated articles: Dislocations and the Mechanical Properties of Crystals (IL, Moscow, 1960) pp. 5-18.

102. V. L. Indenbom and G. E. Tomilovskii, "Microstructure of stresses in slip lines and dislocations," Doklady Akad. Nauk SSSR 123, 673-676 (1958).

103. V. L. Indenbom, V. I. Nikitenko, and L. S. Milevskii, "Observations of internal stresses around dislocations in silicon," Doklady Akad. Nauk SSSR (1961); "Polarization-optical analysis of dislocational structure of crystals," Fiz. tverd. tela (in press).

104. V. N. Rozhanskii and V. L. Indenbom, "Accumulations of dislocations in crystals containing impurities," Doklady Akad. Nauk SSSR 136, 1331-1334 (1960).

105. Fracture (Proc. Internat. Conf., Massachusetts, 1959).

106. V. L. Indenbom, "Criteria of destruction from the viewpoint of the dislocation theory of strength," Fiz. tverd. tela 3, 2071-2079 (1961).

107. V. L. Indenbom and A. N. Orlov, "Physical theory of plasticity and strength," Usp. fiz. nauk (in press).

108. M. V. Klassen-Neklyudova, G. E. Tomilovskii, and M. A. Chernysheva, "Process of formation of kinks," Kristallografiya 8, 646-649 (1960).

109. V. L. Indenbom and A. A. Urusovskaya, "What is meant by incoherent twins?" Kristallografiya 4, 90-98 (1959).

110. M. V. Klassen-Neklyudova, Mechanical Twinning of Crystals [in Russian] (Izd. AN SSSR, M., 1960).

111. V. L. Indenbom, "Reciprocity theorem and effect functions for the tensor of dislocation density and the tensor of incompatibility of deformation," Doklady Akad. Nauk SSSR 128, 906-909 (1959).

V. F. MIUSKOV

EFFECT OF GRAIN DISORIENTATION ANGLES ON THE STRUCTURE AND PROPERTIES OF INTERCRYSTALLINE BOUNDARIES

The areas of contact between grains in polycrystalline bodies are usually called intercrystalline layers or grain boundaries. The discovery of polygo-nization [1, 2, 3, 4], which leads to the formation of a substructure in grains of polycrystals or in a monocrystal as a whole, led to the concept of sub-boundaries— areas of contact between subgrains. In studying the x-ray reflections from crystals many years ago, Darwin introduced the concept of mosaic-block structure in crystals [5]. The separation of boundaries between blocks in the crystal or in the grain of a polycrystal are usually called block boundaries. The difference in these types of boundaries is determined by the degree of disorientation and the size of grains, subgrains, and mosaic blocks.

The sizes of the structural elements in polycrystalline bodies and the angles of their mutual orientation are given in Table 1.

TABLE 1

Structural element	Dimensions, mm	Degree of mutual orientation
Mosaic blocks	10^{-5}-10^{-3}	$5''$-$5'$
Subgrains	10^{-3}-1	$5'$-$2°$
Grains of polycrys-tals	10^{-3} and higher	$1°$-$90°$

For a long time the intergranular layers and subboundaries were separate objects of investigation, although it was indicated that they have a common nature [3]. It is only recently that new dislocation concepts of the structure of intercrystalline boundaries gave a rational foundation for simultaneous in-vestigations of the structure of separation boundaries between blocks, sub-grains, and grains of polycrystals. Simultaneous analysis of the structure and boundaries of blocks, subgrains, and grain boundaries makes it possible to generalize this problem and to resolve a certain number of contradictions which have existed in the past. The structure and the properties of intergranular boundaries is one of the main problems of the strength and heat resistance of materials, which are of particular value in the technology of the present and future. The structure of intergranular boundaries is at present the main problem in the dislocation theory. Intergranular boundaries are very convenient for theoretical and experimental investigations from the viewpoint of the dislocation theory. In some cases the dislocation model of the boundaries makes it possible to derive very important relationships which can be checked experimentally. The problem as a whole (including block boundaries, subboundaries, and grain boundaries) is directly related to the problem of the structure of a real solid. We have not attempted to cover all the existing experimental data on grain boundaries. The object of this article is a review of the most recent experimental data and concepts of the structure and properties of intergranular boundaries based on dislocational models of the structure of these boundaries. We shall discuss only those investigations in which the structure and properties of the boundaries are viewed as a function of the disorientation angle of grains.

EFFECT OF THE DISORIENTATION ANGLE ON THE STRUCTURE AND PROPERTIES OF GRAIN BOUNDARIES

As early as 1929, Hargreaves and Hills [6] adhered to the theory of "transitional layers" and noted that the structure of boundaries must depend on the orientation of the lattices of adjacent grains, i.e., on the angle of disorientation. These authors considered the simplest cubic lattices and derived quantitative relationships between the assumed structure of a transitional zone (boundary) and the rotation angle of adjacent grains. Apparently this was the first indication that if the material of the boundary is not amorphous but has a structure, then the structure must be determined by the angles of mutual rotation of the crystal lattices of grains. Later, Chalmers [7] developed this hypothesis and concluded that if the structure of the boundary depends on the angle of mutual orientation of grains then the different properties of the boundaries must also depend on these angles. However, the first experiments did not confirm this hypothesis. Thus, in studying the lowering of the melting point of the boundaries of a tin bicrystal Chalmers [8] showed that the decrease of the melting point did not depend on the angle of disorientation of grains. Nor did he find later, in 1949 [9], any relationship between the energy of surface tension of boundaries and disorientation angles from 11 to 80°.

Nevertheless, the idea that there is a relationship between the structure and properties of boundaries and the angle of disorientation has progressed: the early experimental and theoretical work [10, 11] was reduced to enumeration and discussion of different properties of the boundaries in the light of the hypothesis that the boundary consists of an amorphous cement or has a transitional structure.

The failure of Chalmers' first experiments to determine the relationship between the properties of the boundaries and the disorientation angles of grains was explained by the lack of any quantitative theoretical data and the lack of a model of the structure and properties of intergranular boundaries.

DISLOCATION MODELS OF BURGERS-BRAGG BOUNDARIES

In 1940, Burgers and Bragg [12] proposed a dislocational model of the boundaries of blocks with small disorientation angles. In this model the boundary between two grains is represented by a row of edge or screw dislocations. Edge dislocations are represented as a line terminating the atomic plane. Figure 1 shows the scheme of the distribution of atoms around an edge dislocation in a plane perpendicular to a line of dislocations in a simple cubic lattice. The figure shows that each dislocation induces local distortion of the lattice and is a source of internal stresses. The field of stresses around dislocations can usually be divided into two regions. The region with a radius R_0 includes significant lattice defects and the dislocation line itself, which usually is called the dislocation nucleus or channel. Outside the dislocation nucleus there is a region, R of distortion of the lattice whose elastic deformation satisfies Hooke's Law.

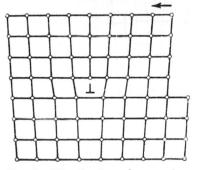

Fig. 1. Distribution of atoms in the lattice around an edge dislocation. The arrow indicates the direction of the slip vector.

Figure 1 shows that the region adjacent to an edge dislocation above the slip plane is in a state of compression due to the presence of an additional atomic plane. An analogous region below the slip plane is in a state of tension. If these dislocations are sufficiently mobile as the result of some condition (increased temperature, or applied stress— creep) then the moving dislocations arrange themselves in a row so that the compressed region of one dislocation will coincide with the elongated region of the other dislocation. Such superimposition of elongated and compressed regions leads to the decrease of the elastic energy of dislocations. Such rows of dislocations have a considerable stability since the energy of dislocation decreases during their formation. Furthermore, during the formation of a row of dislocations the distortions of the lattice due to single dislocations add together in such a way that the macroscopic distortion of the lattice becomes a sum of separate local distortions of the lattice around dislocations. Cahn [4] has used the concept of the arrangement of dislocations in a row in discussing the experimental results on the polygonization of deformed monocrystals of different metals. He showed that dislocation "walls" or subboundaries of a polygonized structure are always perpendicular to the slip line. Dislocations must move only a certain distance in their own slip planes to locate themselves one above the other and thus form a vertical row (Fig. 2b). However, to form such a vertical row

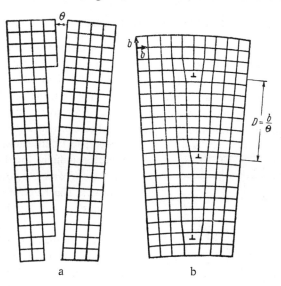

Fig. 2. Distribution of dislocations in a simple boundary with one degree of freedom, θ.

part of the dislocations from the horizontal row which is formed in the slip plane will move in the direction perpendicular to the slip plane so that the distances between dislocations in a new row are identical. During vertical displacement ("crawling over") of dislocations from one plane of the crystal onto another there must be a transfer of mass, i.e., diffusion. When a dislocation passes from a horizontal row to a vertical row its elastic energy decreases. A crystal or a crystallite is transformed from an elastically bent aggregate into an aggregate formed from discretely rotated blocks divided by dislocation walls. A row of dislocations and their accompanying distortion of the lattice insures the rotation of neighboring fragments (block subgrains) at a certain angle θ, which as can be seen from Fig. 3, is equal to

$$\theta = \frac{b}{D},$$

(1)

where D is the distance between dislocations and b is Burgers vector. This formula is valid only for small angles of rotation. For large angles of rotation we have

$$\frac{b}{D} = 2 \sin \frac{\theta}{2}.$$

(2)

Figure 3 shows that with the exception of those places where the atomic plane ends on the boundary the sites of the lattice can approach each other on the grain boundary because of the elastic deformation.

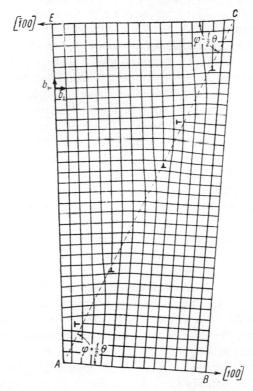

Fig. 3. Distribution of dislocations in a boundary with two degrees of freedom, θ and φ.

Analysis of grain boundaries with large disorientation angles is more complicated and the dislocation theory can no longer give a unique description since in this case the dislocations are so close to each other that their individual particularities are obscured. Consequently one must begin the investigation on boundaries with small disorientation angles since the theory and the experiments referring to this case can be interpreted much more precisely.

Knowledge of the structure and properties of boundaries with small angles will in turn facilitate the solution of more complex problems concerning grain boundaries with large angles. In fact, the discovery of the phenomenon of polygonization and the study of the structure and properties of sub-boundaries helped considerably to widen our concept of intercrystalline boundaries.

Relationships (1) and (2) give a physical meaning to the dependence of the structure and properties of the boundary on the disorientation angle of grains in a dislocation model. It turns out that the dislocation density depends on the disorientation angle of grains. This, one of the most important assumptions of the dislocation theory of grain boundaries, can be checked experimentally. In view of this we shall briefly consider what degrees of freedom the grain boundary can have and what disorientation angles can exist between grains (assuming plane boundaries, for simplicity). In the general case the boundaries have five degrees of freedom characterized by the rotation angles: three angles of rotation of grains around three mutually perpendicular axes of coordinates, θ_1, θ_2, θ_3, and two angles of rotation of the plane boundary itself, φ_1, φ_2. In the latter case only rotation around the two axes in the plane of the boundary has real significance; the rotation of the plane of the boundary around the axis perpendicular to it has no meaning, as can easily be seen.

Boundaries with five degrees of freedom represent a very complex case. Let us first investigate some simple types of boundaries which have a theoretical and experimental interest.

The simplest is the boundary with one degree of freedom, which is determined by only one angle of rotation of grains θ with respect to one axis of coordinates while the angle of rotation with respect to the other axis will be zero. Furthermore, the plane of the boundary is located symmetrically with respect to the grains cutting the disorientation angle θ into two. Such a boundary is called a simple rotational or symmetrical boundary. Figure 2b represents a simple boundary formed by grains of a simple cubic lattice. Here θ is the angle formed by two <100> crystallographic directions of each grain.

Figure 3 represents a more complex case, where the plane of the boundary deviates from the symmetric position by an angle φ. Such a boundary has two degrees of freedom, θ and φ, and is sometimes called the asymmetric boundary. Furthermore, it is important to distinguish the rotational boundary from the twist boundary. The rotational boundary represents the case where the axes of mutual rotation of grains are in a plane of the boundary itself. Then the angle of rotation around the axis perpendicular to the plane of the boundary is zero. A simple boundary (Fig. 2) and a boundary with two degrees of freedom (Fig. 3) are purely rotational boundaries and are composed of a row of edge dislocations. Twist boundaries are formed when the grains are rotated only around axes perpendicular to the plane of the boundary. In this case the boundary is represented by some crystallographic plane common to both grains. Twist boundaries are composed of rows of screw dislocations. Read [31] proposed a general mathematical formula for the model of grain boundaries; using this formula one can construct all the possible types of boundaries composed of different types of dislocations. Amelinckx [72] has investigated the geometry of dislocation nets and fold boundaries for different crystallographic structures using this formula. This study and a number of others are discussed in detail by Amelinckx and Dekeyser [73].

ENERGY OF DISLOCATION ROWS—GRAIN BOUNDARIES

As we have indicated, there is a field of distortions around each dislocation. This field can be divided into two zones according to the character of the distortions: 1) the nucleus, or the distorted material; 2) the zone of stressed undistorted material. The dislocation nucleus consists of material with large distortions of the atomic lattice where the atoms are considerably displaced from their normal positions. At the present time it is difficult to calculate these distortions. The distortions in the zone of distorted material satisfy Hooke's Law. Therefore on the basis of the theory of elasticity one can calculate the stresses and energy created by one dislocation as well as by a series of dislocations at such distances from each other that their nuclei do not interact.

Stresses for linear dislocations in isotropic and anisotropic media were calculated. The value of tangential stresses around the dislocations in undistorted material is given by an infinite series of the type

$$\tau = \tau_0 \frac{b}{R} + \tau_1 \frac{b^2}{R^2} + \tau_2 \frac{b^3}{R^3} + \cdots + \tau_n \frac{b^{n+1}}{R^{n+1}}, \tag{3}$$

where R is the distance from the center of the dislocation. For sufficiently large distances from the dislocation center, R, where R/b is sufficiently large, all the terms of orders beyond the first can be neglected and we obtain an approximate solution

$$\tau = \tau_0 \frac{b}{R}. \tag{4}$$

The value of τ_0 will have great significance in what follows. In the case of an isotropic body whose elastic behavior is determined by two constants, G (shear modulus) and ν (Poisson coefficient), the value τ_0 is of the form

$$\tau_0 = \frac{G}{2\pi (1 - \nu)}. \tag{5}$$

In the case of an anisotropic material τ_0 has a rather complex form; for example, for cubic crystals it is

$$\tau_0 = \frac{1}{2\pi}\left[c_{44}\cos^2\alpha + \sin^2\alpha\,(c_{11} + c_{12})\sqrt{\frac{c_{44}\,(c_{11} - c_{12})}{c_{11}\,(c_{11} + c_{12} + 2c_{44})}}\right], \quad (6)$$

where c_{11}, c_{12}, and c_{44} are the elastic constants of the material and α is the angle between the Burgers vector and the direction of dislocations. For a face-centered cubic lattice α can assume values of $\pi/2$ or $\pi/4$ while for a body-centered cubic lattice $\alpha = \arccos(\frac{1}{3})^{\frac{1}{2}}$. We are most interested in the energy of the grain boundary as a function of the angle of disorientation i.e., the energy of vertical rows of dislocations. It must be noted that the energy of the grain boundaries is a surface energy. Read and Shockley [13] gave a rather strict solution for the relationship between the energy of the simple boundary and the angle of disorientation by summing the stress fields of dislocations located in the boundary. The expression which they obtained for a specific surface energy of a simple grain boundary as a function of the angle θ has the form

$$E = E_0\theta\,(A - \ln\theta) \quad (7)$$

and

$$E_0 = \frac{b\tau_0}{2}, \quad (8)$$

where A is the constant of integration, which depends on the unknown energy of displacements of atoms in a poor material of the nucleus. The value of E_0 depends on the elastic properties of the material by way of τ_0. In Eq.(7) both terms have a simple physical meaning. Aside from the unknown displacement energy of atoms in the nucleus (distorted material) which is determined by the factor A, the first term of this equation, $E_0\theta A$, also contains E_0 and θ, which express the elastic energy of dislocations. The value of the elastic energy is proportional to the dislocation density, $1/D = \theta/b$, i.e., this term indicates a proportional increase of the energy of the boundary of the row with increasing angles of rotation.

The second term of the expression, $\theta E_0 \ln\theta$, is directly related to the elastic energy of dislocation and expresses the fact that the energy of dislocation rows decreases with increasing angles as the result of the superimposition of fields of neighboring dislocations of different signs, which leads to the relief of stress between them, since the field of compression stresses of one dislocation is superimposed on a field of tensile stresses of the neighboring dislocation.

Further, Read and Shockley [13] showed that for a boundary with two degrees of freedom the form of the dependence of the specific surface energy of boundaries on the disorientation angles of grains is preserved, but in this case E_0 and A depend on φ:

$$E_0 = \frac{\tau_0 b}{2} (\cos \varphi + \sin \varphi), \tag{9}$$

$$A = A_0 - \frac{\sin 2\varphi}{2} - \frac{\sin \varphi \cdot \ln \sin \varphi + \cos \varphi \cdot \ln \cos \varphi}{\sin \varphi + \cos \varphi}. \tag{10}$$

For $\varphi = 0$ or $\pi/2$ these expressions take a value equal to the values for simple boundaries.

Expression (8) has a maximum for the abscissa, $\theta_m = e^{A-1}$ or $\ln \theta_m = A-1$, i.e., the position of the maximum depends only on the energy of the nucleus of the dislocation. The value of the energy, E_m at the maximum is equal to

$$E_m = E_0 \theta_m (A - A + 1) = E_0 \theta_m. \tag{11}$$

Expression (8) can be written in dimensionless units by using the relationship ([12])

$$\frac{E}{E_m} = \frac{\theta}{\theta_m} \left(1 - \ln \frac{\theta}{\theta_m} \right). \tag{12}$$

Furthermore expression (8) can also be represented as

$$\frac{E}{\theta} = E_0 A - E_0 \ln \theta. \tag{13}$$

Plotted in E/θ versus $\ln \theta$ coordinates, it is a straight line with slope E_0.

These last two equations allow us to check the dislocation model of the grain boundaries directly by experiments. Equation (12), which represents this relationship in dimensionless units, must be satisfied simultaneously for a series of materials (e.g., for cubic metals), while expression (13) allows us to determine E_0 on the basis of the data on the variation of the absolute energy of the boundary as a function of the angle, and then E_0 can be compared with the value calculated from the elastic constants [see Eq. (6)].

MODEL OF BOUNDARIES WITH LARGE ANGLES (MOTT MODEL)

As we have said, the dislocation model of Burgers-Bragg [12] in the form in which it has been developed today is applicable only to boundaries with a small angle of the relative rotation of the grains and of orientation of the boundary. The main difficulty of the theory of boundaries with large angles is the fact that in these boundaries the dislocations are so close to each other that they cannot be regarded as individual defects.

Recently Mott [15] examined the results obtained by Ke [14] concerning the "viscous" behavior of grain boundaries under the effect of very small loads and proposed a model for boundaries with large angles which makes it possible to explain the viscous behavior of boundaries obtained in experiments concerning imperfect elasticity under small stresses from the viewpoint of the theory of the transition zone. Although the starting point of this model is the mechanical behavior of boundaries with large angles, nevertheless it was successfully applied for a qualitative explanation of other phenomena, e.g., preferential diffusion along the grain boundary. Therefore it seems necessary to discuss the Mott model in more detail.

Mott assumed that the boundaries of grains with large angles (more than 10-15°) are composed of islands of good correspondence divided by regions with greatly distorted structures.

According to Mott the elementary process of slip along the boundary, i.e., the relative motion of two crystalline surfaces, occurs when the atoms around each of the islands of good correspondence become disordered ("melted over"). The free energy F, necessary for the disorganization process will decrease with increasing temperature and at the melting point will be zero. At absolute zero it will be equal to nL, where L is latent heat of fusion per atom and n is the number of atoms in the island with good correspondence. Mott assumed that the temperature dependence of F is given by

$$F = nL\left(1 - \frac{T}{T_m}\right),$$
(14)

where T_m is the melting point. The shear stress τ applied to the boundary will increase the rate of flow in the direction of the action of the applied stress or decrease it in the reverse direction by $+\tau\, na\omega$, where a is the value of the elementary slip or the distance between two neighboring equilibrium states and ω is the area per unit atom.

On the basis of this assumption the rate of slip will have the form

$$v = 2\nu\alpha \exp\left\{-\frac{nL\left(1 - \frac{T}{T_m}\right)}{kT}\right\} \operatorname{sh}\frac{an\omega\tau}{2kT},$$

where ν is the frequency of vibration of the atom. For small values of τ it will have the form

$$v = \frac{\nu a^2 n\omega\tau}{kT} \exp\left(\frac{nL}{kT_m}\right)\exp\left(-\frac{nL}{kT}\right).$$

This expression is a much better description of the experimental fact than if, by analogous reasoning, one operates not with L and T_m but only with activation energy u. It should be noted that in Mott's theory n has a very large value since it is related to the structure of the boundary itself and it can be calculated by the equation, using the experimental values of F and L. Mott's calculation [15] gave n = 14. He assumed that for boundaries with large angles n is constant, when it appears more probable than n must depend on the

disorientation angles of grains. Mott's island model was later used by
Smoluchowskii to explain the phenomenon of preferential diffusion and the
anisotropy of diffusion along the grain boundaries. The scheme of the island
model of the grain boundaries worked out in greater detail by Smoluchowskii
is given below, in Fig. 17. Friedel and his co-workers [74] have calculated
the energies of boundaries with large disorientation angles using a somewhat
improved Mott island model [36]. The islands of strong discontinuities of the
crystal lattice or accumulations of vacancies they consider to be "pseudodislocations" distributed regularly along the grain boundary. For cubic metal
crystals (Cu, Fe, Al) they have calculated the twin boundary common to two
grains located in the planes symmetrical with respect to both grains. In cal-
culating the deformation energy they took into account the energy of the distortion and the rupture of bonds in compressed and elongated regions of
"pseudodislocations." The calculated values of energy for iron containing
silicon were in good agreement with those obtained experimentally for medium and large angles, where the agreement is better than with the Read and
Shockley relationship [Eq. (12)]. There is also agreement for small disorientation angles of grains.

The concepts of the structure of boundaries with small and large angles
described here were put in their final form in 1948-1950. As far as we know
there have been no significant new concepts of the structure of intergranular
boundaries.

PROPERTIES OF GRAIN BOUNDARIES. EFFECT OF THE DISORIENTATION ANGLE OF GRAINS ON THE PROPERTIES OF BOUNDARIES

In this section we shall discuss the most interesting experimental investigations (from our point of view) which have been made since 1948. We
shall consider here only the data on the properties of the boundaries which to
some degree make it possible to draw conclusions one way or another on the
structure of the grain boundaries. The most important properties are the chemical properties of the boundary (etchability), surface tension, preferential diffusion along the boundary, and the motion of boundaries with small angles under the effect of applied stress.

CHEMICAL PROPERTIES OF THE BOUNDARIES

It has been known for a long time that the intergranular boundary is
etched very easily, i.e., the substance of the interlayer dissolves much faster
than the crystal grain. During etching it is usually the grain boundaries which
are put in evidence first while the structural characteristics of the grain itself
are revealed much later.

The first experiments concerning the preferential etchability of intergranular boundaries were made with alloys or technically pure metals containing considerable quantities of impurities. These experiments did not lead to
any conclusion on the structure of the intercrystalline boundary since the
greater solubility of the intergranular layer could be due either to its structure
differing from the structure of the grain or to the presence of impurities in this
boundary.

The first investigation concerning the etching of intergranular boundaries
in very pure aluminum were apparently made by Lacombe and Yannaquis [16].
These authors investigated intercrystalline corrosion of aluminum 99.993%
and 99.998% pure in a 10% HCl aqueous solution. The latter sample contained
0.0002% Fe, 0.0009% Si, 0.0003% Cu, and traces of zinc were found by spectral
analysis. By comparing their results with those obtained by Rohrmann in 1934
[17], who experimented with aluminum 99.95% pure, the authors concluded
that the rate of etching of the grain boundaries does not decrease with increasing purity of the material and that the impurities play only a secondary role in
the increased etchability of intergranular boundaries.

It was very often suggested that the significant role in the increased solubility of boundaries is due to insoluble impurities whose role increases in
quenched metal, when the soluble impurities are transferred into the solid
solution as the result of quenching. However, in the experijents of Lacombe
and Yannaquis [16] with 99.998% pure aluminum the aluminum contained
only 0.002% insoluble iron. It is difficult to imagine that such a small
amount of iron would be sufficient to create around the grain a continuous
envelope which would appear on etching as a continuous boundary. Furthermore, the authors [16] established qualitatively that not all the grain boundaries are etched at the same rate; the etching rate increases with the difference in the orientation of two neighboring grains. High resistance to etching
was observed not only in the case of grains with small disorientation angles
but also in the case of twin boundaries and boundaries corresponding to grains
whose mutual orientation was close to that of the twin.

In these experiments the etching rate was determined qualitatively while
the disorientation of grains was studied by use of Laue diagrams. Later on [3]
the conclusions concerning the relationship between the etching rate of grain
boundaries and the disorientation angles were reconfirmed, and in this case
the orientation was determined by etching figures obtained by selective etching of aluminum.

The decreased solubility of grain boundaries with small disorientation
angles was at first a difficulty in the study of these boundaries. The attempt
to put these boundaries in evidence by using ordinary reagents for a longer
time did not lead to the desired result since in this case the grains themselves
were strongly etched, leading to fuzziness in the whole picture. As a result

there were attempts to develop reagents capable of selective etching. Parallel to these studies and independent of them there were developed methods of selective etching to put in evidence the etching figures on the surfaces of crystals. Thus, Mahl and Stranski [18] examined etching figures in aluminum with the electron microscope. They obtained these figures by using a complex mixture of acids and gaseous chlorides. Many investigators were occupied in the development of such selective etching agents. Among these we should note Tucker [19], Barrett and Levenson [20], and Lacombe and Beaujard [3], who developed and successfully applied selective etching agents to reveal etching figures on the surface of large aluminum crystals. It turned out that these reagents gave unexpectedly good results in the case of boundaries with small angles, i.e., subboundaries. This made it possible to assume that the defects within the grain and in the grain boundaries are apparently of the same nature. Thus the increased local solubility of the metal responsible for etch pits as well as the increased solubility of the boundaries is due to the presence of activating local defects which accumulate on the boundary and form definite rows, while on the surface of the crystal they are localized in separate areas.

It also turned out that the subboundaries appear on etching not as continuous lines as is the case in grain boundaries but are composed of separate etch pits (Fig. 4). The authors also found that the density of the etching figures on the subboundaries of aluminum depend on the mutual orientations of subgrains. In a large number of experiments they showed that the difference in the distribution of etching figures on the boundaries in the case of selective etching is not the result of particular experimental conditions. The experiments with samples whose thickness was that of the grain are particularly convincing. Under these conditions the distribution of the contours of etching nets on both sides of the sample turns out to be the same (Fig. 4a, b). The fact that the net of etching figures is the same on both sides of the sample shows that they are the result of defects within the sample which extend to the surface.

Read and Shockley [21], using formula (1), which correlates the density or the distance between dislocations in the boundary with the angle of rotation of grains forming this boundary, found in 1949 that if the angle between grains is of the order of one degree then the distance between dislocations must be about 100 A. This distance is within the resolving power of the electron microscope. For a subgrain with an angle of 20' the distance between defects will be larger than $3 \cdot 10^{-3}$ mm, which can be resolved with an optical microscope. Therefore Read and Shockley considered discrete selective etching of subgrains described by Lacombe and Beaujard [3, 16] as a fine dislocation structure of the subboundary, represented in Fig. 4a, b. The average distance calculated on the basis of this etching pattern turned out to be

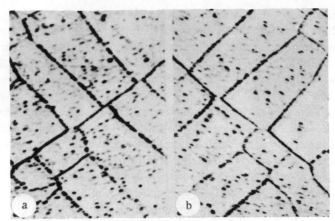

Fig. 4. Discrete etching of subboundaries. a) One side;
b) the other side of the same plate whose thickness was
equal to one grain.

$D = 3 \cdot 10^{-2}$ mm. According to Eq. (1) the value of θ turned out to be equal
to 10^{-4} radian while the direct x-ray measurements by Lacombe and Beaujard
gave a value of $1.3 \cdot 10^{-3}$ for this angle, i.e., thirteen times higher. This
rather large discrepancy can be explained by the fact that either the x-ray
photographs and photomicrographs were obtained from different samples or
were due to other causes which we shall discuss later in this section.

The best confirmation of the correctness of the dislocational model of
subboundaries came from Vogel and his co-workers [22, 23], who thoroughly
checked relationship (1). We shall describe this classical work in detail.

Germanium monocrystals were prepared by zone melting; the direction
of growth was [100]. The subboundaries were observed on the (100) face located in the ($0\bar{1}1$) plane in the zone whose axis coincided with the direction
of growth. The samples were cut so as to reveal the (100) and (011) planes.
The planes were polished with a fine corundum powder and etched two minutes in a CP-4 type etching agent (25 ml HNO_3, 15 ml glacial CH_3COOH, and
a few drops of Br_2). Conical etching figures appeared on the surface of the
(100) face but were not visible on the (011) face, which as can be seen, was
perpendicular to the ($0\bar{1}1$) plane of the subboundary and parallel to the direction of dislocation. The (100) plane, with the etch pits located along the subboundary as a series of discrete overlapping conical figures, is represented in
Fig. 5.

The distance between the etch pits along the subboundaries on the (100)
plane was measured with a microscope under a magnification of 1000. The

distance D between etch pits was averaged not only along the whole length of the subboundaries but also along a few cross sections of the crystal. The authors calculated the maximum error of these measurements to be 10%. X-ray measurements of the angle of rotation of two subgrains were made on the (011) plane of the same sample. At the same time the curves of oscillation on both sides of the boundary were determined by measuring the variation of intensity of the reflection from the (220) plane, using the ionization method. The angle of rotation of two subgrains was determined by the difference of the angular positions of the corresponding maxima of the two curves. The precision of the determination of the disorientation angles was considered by the authors to be equal to ±2.5". The results of metallographic and x-ray measurements are given in Fig. 6. The solid line was calculated by formula (1); the value of Burgers vector was taken as $b = \frac{1}{2}[110] = \frac{1}{2}a\sqrt{2} = 4.0$ A. Figure 6 shows that the theoretical and experimental data are in good agreement. This proves that the etching figures revealed on the boundary are actually defects of the edge dislocation type which are parallel to the axis of growth, [100]. If the defects extend only to the surface of the (100) face, they cannot be revealed in the case of simple tilt boundaries on planes parallel to the direction of dislocations. These experiments convincingly demonstrate the correctness of the dislocation model of subboundaries.

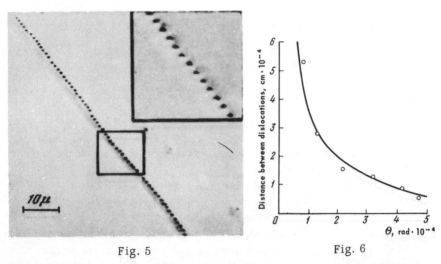

Fig. 5   Fig. 6

Fig. 5. Etching figures on subboundaries of germanium monocrystals.
Fig. 6. Variation of the distance between dislocations in subboundaries as a function of the disorientation angle of grains. The curve represents the equation $D = \dfrac{4.0 \cdot 10^{-8}}{\theta}$.

There exists a somewhat simplified method for strict experimental checking of quantitative relationships derived from the dislocation model of the
subboundaries of crystals with simple cubic lattices. The possibility of such
checking is based on the use of the so-called L and T type intersections of
subboundaries on the planes to which the dislocations extend. Such intersections make it possible to check the theory without measuring the difference
in the orientation of subgrains. The sum of the angles θ of subgrains meeting
in L and T intersections must be equal to zero, i.e., $\Sigma\theta_1 = 0$. This is the consequence of simple geometric relationships valid only for simple cubic crystals. Therefore there must exist a simple relationship between the densities
of dislocations in branches of L and T intersections. If purely tilt boundaries
change their orientations then the dislocation densities must change. Thus if
ρ is the dislocation density in a symmetrical boundary and ρ' the dislocation
density in an asymmetrical boundary which makes an angle φ with the direction of the symmetrical boundary, then we must have a simple relationship

$$\rho' = \rho\,(\sin\varphi + \cos\varphi). \tag{15}$$

If in the general case there exist three asymmetrical purely tilt boundaries
then we must have a relationship

$$\sum_{i=1}^{3} \frac{\rho_i}{\sin\varphi + \cos\varphi} = 0. \tag{16}$$

For symmetrical subboundaries, i.e., when $\varphi = 0$, one can obtain from (16)
for the T type intersection

$$\rho_A + \rho_B + \rho_C = 0, \tag{17}$$

and for the L type intersection

$$\rho_A + \rho_B = 0, \tag{18}$$

where ρ_A, ρ_B, and ρ_C are the densities of the etching figures in the branches
of intersections A, B, and C, respectively.

Amelinckx [24] has checked these relationships in rock salt crystals. He
measured the density of etch pits on the L and T intersections of subboundaries. Table 2, taken from his work, shows the results of measurements of
density of etch pits for the BP, AP, and CP branches, which intersect at point P.

In the second column of the table $\rho_{observed}$ represents the number of
dislocations in the boundary over a distance of 100 μ. The fourth column of
the table represents the calculated values of the density, i.e., $\rho/(\cos\varphi + \sin\varphi)$.
The sum of the $\rho_{CP} + \rho_{AP}$ for two branches, CP and AP, is equal to the density, $\rho_{calculated}$, for the BP branch, namely, $28.5 + 17.9 = 46.4$, which is
very close to the value of the BP branch, 46.7.

TABLE 2

Branch	$P_{observed}$	φ^0	$P_{calculated}$	$D_{observed}$
BP	46.7	1	46.7	2.14
AP	29.9	3	28.5	9.34
CP	23.3	2.2	17.9	4.29

TABLE 3

Intersection number	Length calculated, μ	Number of etch pits over a distance of 100 μ			
		branch A	branch B	branch C	branch B + C
1 av.	50	90	40	52	92
2 av.	155	47	14	30	44
3	10	170	120	50	170
4	40	132	67	72	139
5	60	53	43	7	50
6	80	34	33		
7 av.	50	58	58		
8	123	23	23		

Pfann and Lovell [25] made similar measurements on germanium mono-crystals in the case of purely rotational boundaries, for which relationships (17) and (18) are also valid. The monocrystals were grown in the same way as described by Vogel [23]. The boundaries were observed on the plane of the cube and were located on the (110) planes parallel to the direction of growth, [001]. The results of the investigation are summarized in Table 3.

Figure 7 shows a photograph of a T type intersection of subgrains [25]. Within the limits of the experimental error the value of the sum of densities of etch pits, B + C (the last column of the table), is in very good agreement with the value of the density of etch pits in branch A (third column of the table) for T type intersections. For L type intersections (the last three lines of the table) the value of the density of etch pits in branches A and B are in astonishing agreement. In the latest work on the dislocational structure of germanium monocrystals by the selective etching method, Pfann and Vogel [26] also indicate the astonishingly precise way in which relationships [17] and (18) are satisfied in the case of L and T intersections of boundaries.

To conclude this section let us recall the convincing proof that the figures of selective etching on the subboundaries do in fact correspond to places where separate linear defects have extended to the surface. In a work previously mentioned, Amelinckx [24] first indicated that if a rock salt crystal is cleaved in two, identical nets of etching figures on the subboundaries will appear on both sides of the fresh break even if the surface of each half is etched under somewhat different conditions.

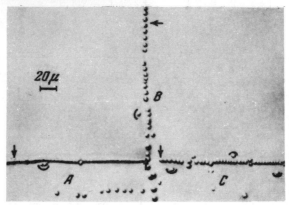

Fig. 7. Etching figures in the branches of subboundaries forming T-shaped intersections.

Gilman and Johnston [27] have shown this same phenomenon in the case of lithium fluoride crystals. They also showed that the number of etching figures on the subboundaries remains constant regardless of the etching time. However, Wyon and Crussard [28] observed a case where the distance between the etching figures did not satisfy Eq. (1) on subboundaries formed as the result of creep.

Most of the experiments described in this section were based on discrete selective etching of subboundaries. However, it should be noted that the mechanism of selective etching still remains unclear in the case of certain substances. In recent years it turned out that the occurrence of etching figures in some crystals depends on the distribution of impurities and the temperature of the heat treatment of the sample.

Wyon and Lacombe [29] have investigated in great detail high purity aluminum with different amounts of iron (from 0.005 to 0.0009%). They concluded that the stresses around the dislocations are much too localized to induce the occurrence of etching figures. These investigators consider, however, that the stresses are sufficient to attract the impurity atoms to a neighboring dislocation. Consequently the etching figures are formed as the result of interactions between dislocations and impurity atoms. Furthermore the etching figures are the consequence of the presence of impurity atoms in more or less dense "atmospheres," while an impurity separated into a microscopic or submicroscopic phases does not create conditions inducing the formation of etching figures.

Suits and Low [30] found that the etching figures (corresponding to dislocations) on the slip lines in a bent monocrystal of silicon containing iron appear only after annealing for 15 min at 160°C. They calculated that for a dislocation to be able to induce an etching figure it is necessary to have about 2000 carbon atoms for each dislocation. When the dislocation density is high (large deformation) etching figures do not occur even after annealing if the number of carbon atoms is insufficient.

The missing number of carbon atoms can be introduced into the deformed material by low temperature carburization at 500°C, after which it becomes possible to put in evidence all the dislocations present in the sample. These authors emphasize the important role of the impurities and the necessity of low temperature activation of the displacement of atoms toward dislocations. On the other hand, Pfann and Vogel [26] and others [14] emphasize the absence of any effect of impurities on the size, shape, and distance between etching figures in germanium crystals. Pfann and Vogel did not find any difference in the etching figures in n- and p-type germanium for either high purity germanium or germanium containing 0.01 at.% impurities of groups III and IV of the periodic table. Pfann and Vogel also affirm that the addition of only a few per cent of some soluble element increases the density of etch pits corresponding to dislocations and the density of etch pits on the subboundaries. Apparently this is the result of a considerable change in the nature of the imperfections of the lattice and not of the capacity of the etching agent to reveal the dislocations. Pfann and Vogel therefore assumed that the soluble elements creating the "atmosphere" around a dislocation do not play any significant role in the formation of etching figures in germanium crystals. This important finding makes it possible to consider as correct the conclusions resulting from the quantitative check of relationship (1) along the L and T intersections which was made by Vogel and others [22, 23] and also by Pfann and Lovell [25].

Furthermore, Gilman and Johnston [27] found that the shape, size and density of etching figures are also independent either of the presence of impurities or preliminary heating of lithium chloride crystals.

Thus, in spite of the fact that the mechanism of selective etching is not yet completely understood, the applications of this method give reproducible results in some crystals and make possible a quantitative investigation of the relationship between dislocation models, while for other crystals only qualitative checks of the dislocation model are possible (e.g., aluminum). Possibly this is the reason that the data obtained by Lacombe and Beaujard [3], which were used by Shockley and Read [21], do not completely satisfy relationship (1).

In summary, we must note that the chemical properties of the boundaries depend on the disorientation angle of adjacent grains. For small disorientation angles forming a simple tilt boundary this dependence is almost identical with Eq. (1), which correlates the dislocation density in the boundary with the disorientation angle of the grains on the basis of the dislocation theory. It is clear that the etching figures of boundaries with small disorientation angles corresponds to the discrete defects of which the boundary is built, at least in the case of simple boundaries.

MOTION OF BOUNDARIES WITH SMALL DISORIENTATION ANGLES UNDER THE EFFECT OF STRESS

Quite early Burgers [12] proposed a dislocation model of block boundaries and predicted the possibility of the motion of such boundaries under the effect of applied stress. Read and Shockley [13] made a detailed investigation of this problem. They showed that there are two different mechanisms of the motion of boundaries with small angles. One of the mechanisms is related to the simultaneous motion of all dislocations of the boundary in their own slip planes. The other mechanism is related to the motion of dislocations in the direction normal to their own slip planes. This latter motion necessitates diffusional displacements of atoms and is related to the "viscous" behavior of the boundaries. It usually occurs at high temperatures while the first mechanism is independent of temperature and stress. The motion of boundaries by either mechanism results in macroscopic deformation. Both mechanisms have been investigated in detail in Read's monograph [31].

Let us briefly examine the possible motions of the simplest boundaries. In a simple boundary (see Fig. 2) composed of one series of edge dislocations the motion of the boundary under the effect of stress occurs as the result of pure slip of dislocations in their own slip planes or as the result of motion perpendicular to the slip planes. In both cases the boundary as a whole is displaced parallel to its original direction.

A more complex case is the motion of boundaries with two degrees of freedom, since in this case the boundary is composed of two series of dislocations located in orthogonal slip planes (see Fig. 3). In the case of pure slip the series of dislocations slipping in its own plane will diverge from a common boundary. However, one particular case is possible: when the angle φ is equal to 45° and the dislocations of the horizontal planes slip to the right while the dislocations in the vertical planes slip downwards by the same distance, then the whole boundary will be displaced to the right by pure slip.

A more general case, the so-called homogeneous motion of boundaries
with two degrees of freedom (Fig. 3), is also possible without their splitting.
But such motion necessitates not only the slipping of dislocations in their own
planes but also a normal displacement of dislocations out of their own planes,
i.e., motion due to diffusion. In this case it is necessary also to account for
the diffusional processes [31].

This more complex motion of boundaries with more complicated structures results in a decrease of the rate of the motion of boundaries which depends on the structure of the boundary itself, i.e., the disorientation angle of
the grains. The rate of motion, which decreases with increasing disorientation angles, must also decrease as the result of the increase of the interactions
between dislocation nuclei.

Thus the dislocation theory predicts the possibility of easy motion of
boundaries, the possibility of their splitting or addition, and the dependence
of the resistance to motion on the disorientation angle of adjacent grains.

Let us pass now to the description of experiments concerning the motion
of boundaries with small angles. The first work in this area was published in
1952 by Washburn and Parker [32]. These authors observed the motion of
boundaries during the bending of bicrystalline samples whose grains were
rotated at an angle of 2° by the preceding deformation. They demonstrated
in principle the possibility of motion of dislocation boundaries under the effect of applied stress. Figure 14 in the second article by Urusovskaya in this
collection (see p. 79) represents the motion of a boundary from the original
position (upper figure) to the left by 0.1 mm (middle figure) and to the right
by 0.4 mm after the change of the direction of stress (lower figure). Li,
Edwards, Washburn, and Parker [33] investigated the motion of boundaries with
small disorientation angles in highly pure zinc (99.99%) in greater detail. Bicrystals with a given orientation of boundaries and angle of the grains were
prepared by preliminary bending of the samples at a given angle and subsequent annealing at 400°C. Under these conditions they obtained grain boundaries whose planes were almost perpendicular to the plane of the base, which
served as the observation plane. The motion of the boundaries was studied
with an optical microscope focused on the basal plane. They used oblique
lighting so that the position of the boundary was determined by the difference
in the brightness of the reflections of two crystals divided by a boundary. The
displacements of boundaries were determined with a precision of ± 0.001 mm.

It was shown that the motion of a boundary occurs as the result of applied
shear stresses whose values are approximately equal to the critical shear stress
for slipping in zinc. As the result of this finding the authors concluded that
the flow limit of the material is determined not by the amount of stress, including the effect of the dislocation source of the Frank-Read type [31], but
by the slip stress of dislocations accumulated in the boundary. The direction

of the motion of the boundary changes with the change in sign of the stress. The value of the inverse critical stress is lower each time than the preceding direct stress. This is apparently a phenomenon analogous to the Bauschinger Effect which, as was shown by rather precise experiments [34] also occurs in monocrystals and can be explained by the fact that the motion of dislocations along the same path but in the reverse direction is easier than the original motion. Under constant stress, the smaller the angle of the boundary the greater the velocity of motion. Of particular interest is the fact that boundaries moving with different velocities overtake each other and form one boundary. This new boundary moves with a lower velocity. The motion of boundaries near areas with a deformed lattice is slowed down. With increasing temperature the velocity of motion increases. These authors calculated that the activation energy of this process in the temperature range of 300-400°C is 21,500 cal/mole. It is very interesting that this value is very close to the value of the activation energy for self-diffusion and creep in zinc crystals. The results obtained have confirmed all the predictions of the dynamic properties made on the basis of the dislocation model of grain boundaries.

In 1954 Bainbridge, Li, and Edwards [35] published an investigation in which they describe the motion of small angle boundaries in zinc at temperatures in a very wide range (from −196 to +400°C). They introduced a number of improvements which ensure that the average shear stresses during motion of the boundaries remain constant. The boundaries move under the effect of applied shear stresses acting in the direction of the Burgers vector (direction of slip) of edge dislocations located in the boundary. As in publication [34], these authors found a relationship between the rate of motion of the boundary and the angle of disorientation, which is represented in Fig. 8. The figure shows that with increasing disorientation angles the rate of motion of the boundary decreases according to a law close to hyperbolic. They found that the character of the motion of the boundary under the effect of stress changes with temperature according to the dislocation hypothesis. With constant applied stress the motion is regular at high temperatures (300-400°C). At the highest temperatures

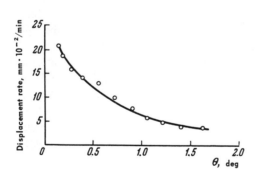

Fig. 8. Variation of the rate of displacement of boundaries with small angles as a function of the disorientation angle of grains.

they could not establish a critical stress, and motion of the boundary occurred at any stress, no matter how small. At +20°C the motion of boundaries was intermittent, the jumps sometimes reaching 0.5 mm. To ensure motion at −196°C it was necessary to increase the stress continuously but the boundaries behaved irregularly, as at room temperature: during motion the boundaries changed their form and sometimes even stopped moving in spite of the continuous increase of the stress. The angle of the boundary remained constant during its motion at high temperature, but at low temperature the angle decreased and the shape of the boundary became more complex due to the interaction with the structural defects encountered during the motion. At high temperatures diffusional processes apparently favor the overcoming of barriers and make it possible for the boundary to move as a whole. At low temperatures the moving parts of the boundary are captured by structural defects, which leads to the distortion of their shape. The negative and positive dislocations captured by the boundaries can be mutually annihilated by the climb mechanism proposed by Mott [36]. Closely spaced boundaries with angles of the same sign at 400°C sometimes unite (according to the rate of motion), i.e., one boundary is formed. The tilt angle of the united boundary turned out to be equal to the sum of the angles of the boundaries united. But the most interesting effect is that such a union of boundaries can occur under the effect of stress at any temperature, including −196°C, when the activation processes are absent. Figure 9 shows how nine closely spaced boundaries unite into one under the effect of stress at −196°C.

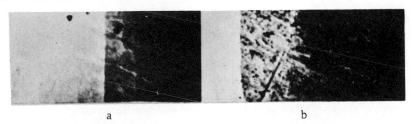

a   b

Fig. 9. Union of nine grain boundaries during motion at −196°C. a) Initial state; b) after union of the boundaries.

The union of boundaries of different signs at low as well as high temperatures was also observed. Earlier [34] it was shown that incompletely united boundaries of the same sign can separate when the sign of the applied stress is changed. But the incompletely united boundaries of different signs do not separate into the original boundaries when the sign of the stress is changed. This can easily be explained by the fact that boundaries with angles of different signs contain dislocations of different signs which can annihilate each other when they come close together. Therefore the process of unification of such grain boundaries under the effect of stress alone without the participation of thermal effects clearly demonstrates the mechanism of formation of substructure during deformation even at low temperatures.

Another conclusion is also very interesting. The results of the experiments on the motion of boundaries in zinc crystals were compared with the results of experiments on slip under the effect of shear stress. It was found that the characteristics of these processes are very similar. From this fact it is possible to assume that it is not the generation of new dislocations but the motion of dislocations through the structural barriers of the crystal which determines the flow limit of the crystal. This assumption is also confirmed by the fact [33] that the critical stress at which motion of boundaries begins is equal to the flow limit of the material.

In summing up all these investigations we can say that the motion of boundaries with small angles and their thermal and athermal behavior has been shown experimentally. The phenomena of combination and decomposition of boundaries with angles of rotation of the same sign and the particular behavior of boundaries with angles of different signs during their combination and decomposition can be explained only on the basis of a model of the boundary constructed of defects which can be algebraically summed. Furthermore, observation of the motion of boundaries with small angles under the effect of stress has led to the discovery of a number of phenomena important in solid state physics and the dislocation theory (the physical meaning of the flow limit, the mechanism of polygonization, etc.).

SURFACE TENSION OF GRAIN BOUNDARIES

The surface tension of grain boundaries has been investigated for a long time. The main object of these investigations has been the effect of the surface tension of grain boundaries on the equilibrium shape of grains of mono- and multiphase polycrystalline metals and alloys. The extensive article by Smith [37] gives a particularly detailed review of these investigations.

As we have seen, it was apparently Chalmers [9] who first attempted to correlate the magnitude of the surface tension of the boundaries with their structure. However, his first attempt to find a correlation between the surface tension of boundaries and the disorientation angle of the grains was not successful because his measurements were limited to large angles, from 10 to 80°, where surface tension is independent of or depends very little on the disorientation angle. It was only after the derivation of Eq. (7), relating the energy of the boundaries and the degree of disorientation, by a mathematical treatment of the Read-Shockley dislocation model [13] that it became clear that small disorientation angles give the most interesting results and that these measurements can be used to check the dislocation model experimentally. In this section we shall describe different experimental methods of checking the

relationship between the energy of the boundary and the disorientation angle
[Eq. (7)].

The surface tension of the grain boundaries can be measured by different
methods. We shall describe the two methods most frequently used.

The Tricrystal Method. If three grains meet on the same edge
then the boundaries between the three grains must reach local equilibrium at
angles determined by the requirement of vector equilibrium of the surface
tension. Figure 10 is a diagram of three boundaries intersecting at one point;
from this figure it follows that

$$E_1 + E_2 \cos \psi_3 + E_3 \cos \psi_2 = 0 \qquad (19a)$$

or

$$E_1/\sin \psi_1 = E_2/\sin \psi_2 = E_3/\sin \psi_3, \qquad (19b)$$

where E_1, E_2, and E_3 are the surface energies of the boundaries and ψ_1, ψ_2,
and ψ_3 are the angles formed by the boundaries. However, these relationships
do not account for the effect of the variation of the orientation of the boundary; if we take this factor into account then terms of type

$$p_i = \frac{1}{E_i} \frac{\partial E_i}{\partial \psi_i} \ (i = 1, 2, 3, \ldots) \qquad (20)$$

will occur in expressions (19a) and (19b), and the value of these terms can be
determined from formula (7), which is valid for small values of θ when
$A \ll - \ln \theta$, and for large values of θ on the basis of experimental data of
a qualitative character. Twin boundaries and boundaries in positions close to
secondary energy minima, where $\partial E_i / \partial \varphi_i$ acquires very large values, require
special treatment, which has been developed by Read and Shockley [13].
They showed that when θ is large there are reasons to assume that the variations of A and E_0 with the orientation of the boundary are mutually compensated and therefore the energy is only slightly dependent on φ. In general
boundaries with small values of θ occur in tricrystals containing two boundaries with large values of θ; other combinations of boundaries are completely impossible or most improbable. Under these conditions $p_2 = p_3 = 0$ and the
relationship between the three angles, ψ_1, ψ_2, and ψ_3 (Fig. 10) is such that
the coefficient at p_1 is close to zero. Thus in practice relationships (19a) and
(19b) are satisfied rather precisely. These expressions are equivalent to the
simple triangle of forces of a mechanical model.

The Bicrystal Method. In this method one measures the angle
between the sides of the "channel" (see Fig. 11) formed by the two grain
boundaries as the result of high temperatures annealing of pure surfaces of
large crystal samples in vacuum or in an inert atmosphere. The atoms on the

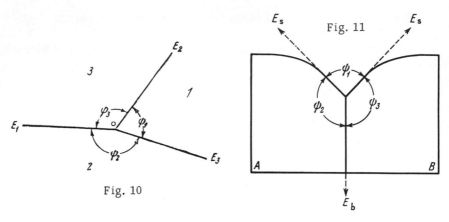

Fig. 10

Fig. 10. Diagram of the positions of three grains (1, 2, 3) in a tricrystal for the deviation of the relationship between the energy and the angles of the boundaries.

Fig. 11. Diagram of the position of grains of a bicrystal (A and B) with a channel along the boundary for the derivation of the relationship between the surface energy of the boundary and the angle formed by the sides of the channel.

grain boundaries migrate away from it along the surface of the grain; such surface migration is possible because the surfaces of the channels approach a shape in equilibrium with the surface of the grain. Figure 11 represents a cross section of a sample composed of two crystals, A and B, with a given orientation. The angle between the sides of the channel ψ_1, has an equilibrium value and the angles ψ_2 and ψ_3 are equal to each other, which occurs when the difference between the surface tensions of the external surfaces of the A and B crystals is vanishingly small. By applying Eq. (19a) to this case we obtain

$$E_b = 2E_s \cos \frac{\psi_1}{2} , \qquad (21)$$

where E_b is the surface tension of the boundary and E_s is the surface tension of the external surfaces of the crystal. If we know E_s, then, from Eq. (21), we can determine the absolute value of the energy of the surface tension of the boundary, and if E_s is not known, then only E_b / E_s can be determined.

When one applies the tricrystal method one usually assumes that one of the energies, say E_1, is constant; this occurs when the orientation of two neighboring grains is fixed while the orientation of the third grain is varied. Thus E_2 / E_1 and E_3 / E_1 and E_3 / E_1 are determined as functions of the disorientation angle.

In order to obtain data comparable with the surface tension it is necessary that the boundary of the samples have an equilibrium shape. In the experimental investigation which we shall consider later the equilibrium state of the boundaries of bicrystals and tricrystals was attained by prolonged annealing at temperatures close to the melting point.

Chalmers and his collaborators [9, 38, 39] measured the surface energies of the grain boundaries of highly pure metals with low melting points: 99.987% tin and 99.999% lead. They prepared tricrystalline samples from a melt by using three seed crystals each of which had a definite orientation. However, tricrystals obtained by this method did not have a common axis of intersection of the three boundaries in the whole series of samples. Dunn[40] proposed a recrystallization method of preparing tricrystalline samples from a thin foil of iron containing silicon (3.35% Si). This method made it possible to ensure that the planes of the boundaries are perpendicular to the surface on which the angles are measured. By this method the orientation of separate grains of tricrystals was ensured by the orientation of the seed crystals. The seed crystals were large grains which resulted from local heating of the foil in three different places. Then the samples were placed in an oven, where the grains grew from the oriented seed crystals. He prepared two series of samples composed of three grains [41, 42]. In the first series the common axis of three grains was the [100] direction, and [110] in the second series. The crystals were annealed 3-4 days at 400°C in an atmosphere of dried hydrogen in order for the three intergranular boundaries to reach an equilibrium position. The angle between the boundaries was measured on micrographs taken at a magnification of 500.

The method of preparing bicrystals used by Greenough and King [43] to determine the surface tension of boundaries does not differ in any way from the Chalmers method for preparing tricrystals from a melt. Two seed crystals with fixed orientations were used to obtain the desired orientations of the two grains grown from the melt. In this case as well as in the crystals obtained by Chalmers, the axes of relative rotation were not strictly constant crystallographic directions but varied somewhat from sample to sample. To obtain a sufficiently deep channel on the boundaries the samples were subjected to prolonged heat treatment at 900°C in an inert atmosphere. In this method the greatest difficulty is to determine precisely the angle formed by the walls of the channel.

The orientation of grains of tricrystals and bicrystals was determined by Laue diagrams with a precision of ±1-2°.

Let us pass now to the problem of the extent to which the experimental data on the relative energy of the boundaries obtained by Dunn and others [41, 42] and Chalmers and Aust [38, 39] are in agreement with the theoretical conclusions drawn from the dislocation model of grain boundaries with

small angles. These data make it possible to check formula (12) directly.
For this purpose all the data on lead, tin, and iron (containing silicon) for the
[110] and [100] series are represented in dimensionless coordinates, E/E_m,
θ/θ_m. In accord with the theory, all four series of data lie along the same
curve (Fig. 12). The experimental points are somewhat scattered and the

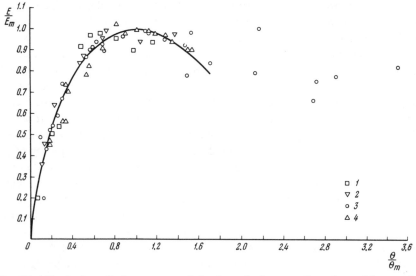

Fig. 12. Variation of the energy of the boundaries as a function of the angles
in dimensionless coordinates E/E_m and θ/θ_m for tin (1), lead (2), and two
series of iron (3, 4).

causes of this scattering are the approximate character of Eq. (12) and the
fact that no account is taken of the effect of the orientation of boundaries φ,
which, according to Eq. (9), leads to the variation of the energy by a factor
of $\sqrt{2}$ when φ varies from 0 to 45°. The values of E_m and θ_m are of some
interest. Since in the experiments described it is the relative energy which
was measured, one cannot determine E_m from the curve. This value will be
determined later from the absolute values of the boundary energy. The value
of θ_m can be determined if the data relative to each series of experiments
is represented in coordinates E/θ and $\ln \theta$; then θ_m is determined by the
intersect on the $\ln \theta$ axis. In this way it turns out that for iron of the [110]
series $\theta_m = 26.6°$, and 29.8° for the [100] series; for tin $\theta_m = 12.2°$, and for
lead $\theta_m = 25.0°$.

Figure 12 shows that the experimental data satisfy the theoretical curve
quite well not only in the case of small angles but surprisingly well in the
case of large angles (25-30°). In the case of large angles the distance be-

tween dislocations D, is equal to b, while for the derivation of formula (8) it
was assumed that D ≫ b. Such good agreement for large angles θ_1 is apparently the result of mutually compensating errors for A and E_0. Even though
in the case of large angles the coincidence of the experimental and theoretical curves may be accidental, and a more precise theory of the model is
needed, Fig. 12 shows convincingly that the energy increases rapidly with θ
and there is a wide maximum as predicted by the model of boundaries.

The ratio of the surface energies of the boundaries of silver bicrystals to
the surface energies of the grain measured by Greenough and King [43] is
shown in $E_b/E_s-\theta$ coordinates in Fig. 13. In this case the scattering of re-

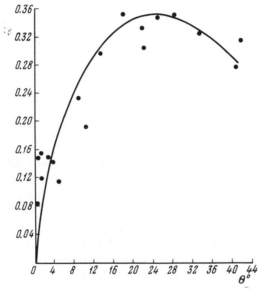

Fig. 13. Variation of the relative energy of the
boundaries E_b/E_s as a function of the disorientation angle θ of boundaries in silver.

sults is somewhat greater, particularly for small angles. However, these data
allow a relatively precise calculation of θ_m, which turns out to be equal to
25°, which is in good agreement with the value of θ_m for lead, which like
silver has a face-centered structure. Thus the experimental data obtained by
Greenough and King also confirm the theoretical curve [Eq. (8)].

The absolute values of the surface energies of silver, which are necessary
to calculate E_b by formula (21), are taken from the work of Sawai and Nishida
[44], where $E_s = 1140$ erg/cm², a value which according to Shuttleworth[45]
must be increased to 1311 erg/cm² because of the effect of transversal bound-

aries of grains present in the sample. Since from the curve of Fig. 13 it follows that $E_m = 0.35 \cdot E_s = 0.35 \cdot 1311 = 460$ erg/cm^2, then $E_0 = E_m/\theta_m = 1050$ erg/cm^2. The theoretical values calculated from the elastic constant by formulas (6) and (9) must be averaged for angles from 0 to $\pi/2$ and must be decreased about 15% to account for the decrease of the elastic constant with an increase of temperature up to 900°C in order to be compared with the experimental values. After all these somewhat arbitrary corrections, which introduce some uncertainties, the calculated value turns out to be $E_0 = 1030$ erg/cm^2, which can be considered to be in good agreement although there are some reasons to expect much less agreement (Read [28], p. 247).

For comparison of the absolute values of the energy of iron containing silicon we used the data of Van Vlack for iron +4% silicon, which were obtained by statistical treatment of a great number of measurements of the energy of the boundary: 780 erg/cm^2 at 1105 °C. This value can be used to determine at least the order of magnitude by way of some not strictly justified assumptions:

1) The boundary may with equal probability form any angle θ and, therefore, statistical analysis gives the average value for energy in terms of θ. In the data obtained by Dunn and others [41, 42] the average value of the energy of grains was 0.85 E_m.

2) The average value of the energy in Dunn's experiments, in which the number of degrees of freedom of the boundary was taken into account, was equal to the average energy of the boundary in Van Vlack's experiments, in which the boundaries had five degrees of freedom. Thus we have $E_m = 760/0.85 = 895$ erg/cm^2, while $E_0 = 895/0.509 = 1720$ erg/cm^2 for the [100] series and $E_0 = 895/0.465 = 1920$ erg/cm^2 for the [110] series. The value of E_0 calculated on the basis of the elastic constants (taking into account their temperature dependence) was, according to Brooks [47]: $E_0 = 2620$ erg/cm^2 for the [100] series and $E_0 = 2910$ erg/cm^2 for the [100] series. But in this case, as one would expect, the discrepancy between the experimental and theoretical values is much greater than in the case of silver due to the fact that the Van Vlack data do not exactly fit our purpose.

Fisher and Dunn [48] made calculations for copper, using statistical results of measurements of the energy of boundaries [31]. They obtained 550 erg/cm^2 as the average of four separate measurements of the surface energy of grain boundaries. These calculations were limited to the calculation of θ_m on the basis of E_0, which was calculated from the elastic constants with a 15% correction for the temperature dependence of the elastic constants. Assuming as before that the average value of $E = 0.85$ E_m, they found $E_m = 650$, the calculated value of E_0 being 1390. Therefore

$$\theta_m = \frac{650}{1390} = 0.465 \text{ rad } = 26.5°,$$

which is in relatively good agreement with the data for lead and silver which, like copper, have face-centered lattices.

All these data on the absolute energies of boundaries are summarized in the table, where the theoretical and experimental data are compared (to the extent possible). One must take into account the fact that there are no experimental data on the relationship between the absolute energy of the boundary and the disorientation angle. The data on absolute energy used in the calculation by Sawai and Nishida [44], Van Vlack [46], and Fisher and Dunn [48] were obtained independently on different samples and under different conditions (different gaseous media and temperatures). One must also keep in mind that the temperature dependence of the elastic moduli was not precisely accounted for. Thus we cannot expect a very precise agreement between the calculated and experimental values (Table 4). Nevertheless, the

TABLE 4. Theoretical and Experimental Data on the Absolute Energy of Grain Boundaries

Material	E_{abs}, erg/cm^2	E_m, erg/cm^2	θ_m exp, deg	θ_m calc, deg	E_0 exp	E_0 calc
Fe 100	760	895	29.8		1720	2620
Fe 110		895	26.6		1920	2910
Ag	1310	460	25		1050	1030
Cu	550	650		25		1390
Pb			25			780
Sn			12.2			

values obtained for the absolute energy and also the relatively good confirmation of Eq. (12) represented by the curve of Fig. 12 allow us to consider that the experimental data on the surface energies of the boundaries are convincing confirmation of the dislocation model of the grain boundaries. At the present time measurements of the angular dependence of the absolute energy of boundaries within a wide range of disorientation angles, including regions of small angles ($\theta < 2$-$3°$), are particularly needed. Also, it is desirable to measure simultaneously the temperature dependence of the elastic constants for the same samples. Only such data can make it possible to draw precise conclusions.

PREFERENTIAL DIFFUSION ALONG THE GRAIN BOUNDARIES

The phenomenon of preferential diffusion along the grain boundaries has been studied for many years. There were reviews of the many publications in this field by Berrer [49] in 1936 and by Klassen-Neklyudova and Kontorova

[10] in 1939. These reviews are essentially comparisons of diffusion rates in monocrystals and polycrystals. It was found that the diffusion coefficient in polycrystals is higher than the diffusion coefficient in monocrystals of the same material. It was also found that the diffusion coefficient of polycrystals increases with decreasing grain size. However, none of these investigations gave any definite information on the structure of the boundary. On the basis of these investigations one can assume only that in the intercrystalline layer the atoms are relatively less densely packed than in the grain. The mathematical analysis of simultaneous diffusion along the grain boundary and within the grain developed by Fisher [50], and later made more precise by Whipple [51], made possible a more precise treatment of the results on diffusion than was possible with the Dushman-Langmuir equation.

Experiments on preferential diffusion along the grain boundaries as a function of the disorientation angles are necessary to construct a model of the grain boundary. Beck, Sperry, and Hu [52] were apparently the first to point out a relationship between the diffusion rate and the disorientation angle; they noted an increase in the displacement rate of grain boundaries during recrystallization with an increase of the disorientation angle. However, it was Smoluchowski and his collaborators who began a systematic investigation of this phenomenon. Achter and Smoluchowski [53] studied the diffusion of silver and copper as a function of the disorientation angle by the metallographic method. To decrease the number of degrees of freedom of the boundary they used copper samples with a columnar structure. The common axis of the columnar crystals coincided (within 7-8°) with the [100] direction. Consequently, twist boundaries with two degrees of freedom, θ and φ, were studied in this experiment. The disorientation angle was determined by x-ray analysis. Since not all the grains had an axis which coincided exactly with the [100] direction, the measurements were made only on grain boundaries whose [100] direction made an angle no greater than 8° with the columnar axis. Diffusion of the vapors of a copper alloy containing 4% silver into columnar copper was investigated at 670, 700, and 725°C.

After diffusional annealing at one of the three temperatures the vapor was directed parallel to the surface of separation, i.e., perpendicular to the direction of perpendicular to the direction of columnar growth, [100]. The depth of penetration of silver in different grains as a function of their disorientation angle was studied at different depths below the original surface. It turned out that preferential diffusion along the boundaries depends greatly on the disorientation angle of neighboring grains. It was also found that preferential diffusion along the grain boundaries occurs only at angles between 20 and 70°. At angles below 20 or above 70° there was no noticeable preferential diffusion of silver along the grain boundaries, i.e., the rates of volume diffusion and of diffusion along the boundaries are equal for small angles of orientation. The curve representing the variation of depth of penetration along the boundary as a function of the disorientation angle has a wide maximum at $\theta = 45°$. The experiments also showed that the angle of orientation φ of the boundary has very little effect on the depth of penetration of silver along the boundaries. However, Dawson [54] has indicated that Achter and Smoluchowski's results could depend on another factor since in systems of the Cu—Ag type the rate of decomposition of the solid solution depends not only on the difference in the orientation of neighboring grains but also on supersaturation.

Apparently as the result of Dawson's remarks, Flanagan and Smoluchowski [55] made a similar investigation of the Cu—Zn system, in which the solid solution does not decompose. Flanagan and Smoluchowski investigated the diffusion of zinc in columnar copper by the method described above. Their observations confirmed their previous results [53] which showed that the rate of intergranular diffusion depends on the disorientation angle. They also found a relationship between the activation energy of the process and the orientation angle of the grains.

The etching method used in the investigations by Smoluchowski and his co-workers to study the depth of penetration is not very sensitive. It should be kept in mind that in diffusion there is not only a flow of atoms of the diffusing element along the boundaries but also into the body of the sample. Therefore, preferential diffusion along the grain boundary can be observed, particularly by the metallographic method, only when $D_B \gg D_L$. Therefore we have grave doubts about the results obtained by Smoluchowski which indicate that there is no preferential diffusion along the grain boundary for small angles, $\theta_s = 10\text{-}20°$ [56].

Smoluchowski [57] later attempted to increase the sensitivity of this method of investigation in order to obtain more precise measurements of the dependence of the depth of penetration and the anisotropy of diffusion along the grain boundaries on the disorientation angle of the grains. In his early work Smoluchowski gives only qualitative data on the anisotropy of diffusion along the grain boundaries and a qualitative explanation of this phenomenon on the basis of a more precise model of the boundary [58]. In more recent work [57] he applied a more sensitive method of investigation of the Cu—Ag system by the use of the radioactive isotope Ag^{110}. Also, the investigation was made not with columnar crystals of copper but with bicrystals prepared by the use of oriented seed crystals. The anisotropy of diffusion was measured in the boundary of a bicrystal in three directions making different angles φ (0, 45, and 90°) with the axis of rotation of the crystals [100]. Three different series of samples were prepared whose original surfaces were at angles of 0, 45, and 90°, respectively, to the [100] direction, the axis of the bicrystal.

The isotope was deposited on the original surface electrolytically, and diffusional annealing at 670°C was continued for 336 hr.

Figure 14 represents the variation of the depth of penetration of Ag^{110} as a function of the disorientation angle θ. The curves relative to 0, 45, and 90° confirm rather precisely the data of the previous work with the Cu— Ag and

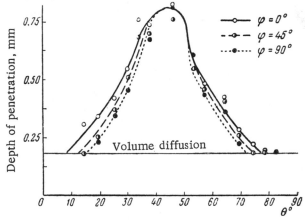

Fig. 14. Variation of the depth of penetration of silver during diffusion along the boundary of copper as a function of the disorientation angle of grains θ for different directions in the plane of the boundary.

Cu— Zn systems investigated by the metallographic method. In this case it was possible to differentiate between the diffusion along the boundaries and within the grain for smaller angles due to the increased sensitivity of the method. The curve for $\varphi = 0°$ intersects the level of volume diffusion for angles as low as $\theta_s \approx 9°$, while earlier it was found that $\theta_s \approx 20°$ for the same system [53]. Consequently the value of θ_s depends on the sensitivity of the method and is not a parameter of the diffusion process.

In Fig. 14 all the curves are asymmetrical with respect to the maximum (at 45°). The asymmetry of the curves apparently indicates the presence of secondary minima of diffusional penetration for angles corresponding to twinning positions or positions with a low density of discontinuities in the atomic structure of the grain boundaries. Figure 14 indicates the existence of a relationship between the depth of penetration of silver and the angle φ, i.e., the existence of anisotropic diffusion. It turns out that the maximum depth of penetration occurs in the [100] direction of the axis of rotation (along the dislocation line) and the minimum in the direction perpendicular to it, i.e., when $\varphi = 90°$. The direction $\varphi = 45°$ assumes an intermediate value. Fig-

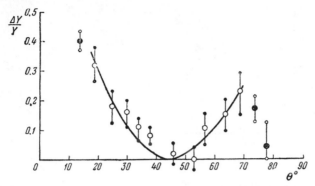

Fig. 15. Variation of the anisotropy of diffusion along the boundaries $\Delta Y/Y$ as a function of the disorientation angle of the grains.

ure 15 represents the value of the anisotropy of diffusion along the boundaries expressed in the form

$$\frac{\Delta Y}{Y} = \frac{Y\,(\varphi = 0^\circ) - Y\,(\varphi = 90^\circ)}{Y\,(\varphi = 0^\circ)} \tag{22}$$

as a function of the disorientation angle, θ. The figure shows that the anisotropy is maximum for small angles. As the angle approaches 45° (whether increasing from small angles or decreasing from larger angles) the anisotropy decreases and when the angle is 45° it is zero, within the limits of precision of the experiment. Thus in purely twist boundaries the diffusion has a directed character which depends on the disorientation angle of neighboring grains.

In a subsequent publication Smoluchowski and Haynes [59] used the radiographic method to investigate the selfdiffusion of iron containing silicon (3.17% Si). They experimented with bicrystals obtained by the method proposed by Dunn and Nonken [60]. The duration of diffusional annealing at 769 and 810°C was chosen so that the depth of penetration was the same at both temperatures. The results of this investigation are given in Fig. 16. The results were the same as in the Cu— Ag system, i.e., for disorientation angles lower than 10° there is no preferential diffusion along the grain boundaries while for angles larger than 10° there is a relationship between depth of penetration and the disorientation angle. The substantially new factor here is good reproducibility of the wide minimum which cannot be related to the secondary minima in the Cu— Ag system [52], where the effect of the minima was only to render the curve asymmetric. Here the presence of the wide minimum is explained by introducing the concept of the density of atomic discontinuities on the grain boundary and is not related to secondary minima on the curve of the energy of the boundaries [Eq. (8)].

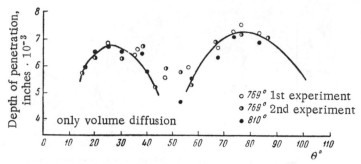

Fig. 16. Variation of the depth of penetration during self-diffusion along the boundaries of bicrystals of iron as a function of the disorientation angle of grains θ.

Therefore, in the whole series of investigations of different systems, Smoluchowski showed: a) that there is a relationship between the rate of preferential diffusion along the boundaries and the disorientation angle of the grains; b) the existence of an anisotropy of diffusion along the boundaries and the dependence of this anisotropy on the disorientation angle; c) the absence of preferential diffusion along the boundaries with small disorientation angles. It should be noted, however, that the last conclusion is still doubtful [56].

On the basis of these data Smoluchowski has constructed a model of the grain boundaries with large disorientation angles which at the same time touches on some details of boundaries with small disorientation angles. The concepts of Smoluchowski on the relationship between the structure of the boundary and the angles θ are shown schematically in Fig. 17. Smoluchowski considers that grain boundaries with a small angle θ are well described by the Burgers-Bragg dislocation model. The grain boundary consists of a series of dislocations (Fig. 17a) but he makes the rather doubtful assertion that diffusion along the nuclei of dislocations through the so-called dislocational tubes is no different from volume diffusion. According to Smoluchowski this is due to the fact that one isolated dislocational tube creates very small pores in the material. With increasing disorientation angles of the grains the distance between dislocations decreases and finally, when the distance between separate dislocations in the boundary is 2-3 interatomic distances, dislocational tubes begin to interact between themselves, creating complexes with a greatly distorted structure of the lattice. These complexes are divided by areas with an undistorted lattice (the blank areas of Fig. 17b). These complexes occur (the islands in Fig. 17b, c), and preferential diffusion at the grains begins, at a definite disorientation angle θ. The rate of preferential diffusion increases with the disorientation angle because the size and number of these complexes increase. This model is the same as the Mott model of the grain boundary

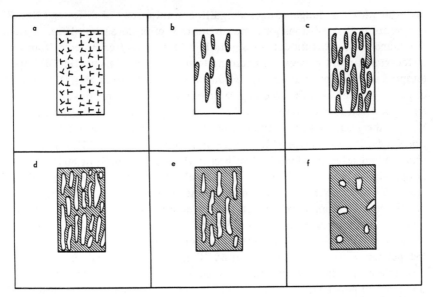

Fig. 17. Diagram of the structure of boundaries as a function of the disorientation angle of grains according to Smoluchowski. a) Simple boundary—edge dislocations $0 < \theta < \sim 7°$; b, c) islands of distorted material (interacting dislocations) in areas of undistorted material $\sim 7° < \theta < \sim 25°$; d, e) islands of undistorted material $\sim 25° < \theta < \sim 40°$; f) isotropic medium $\theta = \sim 45°$.

except that more.details are given: the twist boundaries are represented as rods with an elliptical cross section. The axes of such rods are in the plane of the boundary and parallel to the axis of rotation of the grains. The dislocational model of boundaries, which are built of rows of linear edge dislocations parallel to the axis of rotation of grains, dictates that the rods be parallel to the axis of rotation. The rods of good correspondence (Fig. 17b, c) are oriented in the same way. In such a boundary the diffusion must be anisotropic: the rate of diffusion is maximum in the direction of the axis of the rods and minimum in the direction perpendicular to them, in which direction the rods with a greatly distorted structure are separated by rods with an undistorted structure. The dimensions of rods with a distorted lattice will increase with an increased angle θ, while the dimensions of rods with an undistroted lattice will decrease, which will increase the rate of preferential diffusion along the boundaries and decrease the anisotropy since with increasing disorientation angles not only the dimensions but the number of areas with an undistorted lattice will decrease (Fig. 17d, e).

Finally there comes a moment when the boundary consists of a continuous mass of greatly distorted material containing small islands of undistorted material which no longer create a significant anisotropy of diffusion (Fig. 17f) but create only conditions for a maximum rate of preferential diffusion along the boundaries. This model is essentially a Mott model considered from the dislocational point of view. According to the authors, Vassamillet [61] attempted to calculate the energy of boundaries with large angles. The model of grain boundaries with large angles proposed by Smoluchowski did not arouse any opposition in the literature and is apparently well verified experimentally. But the assumption that there is no diffusion along the isolated dislocational tube led to considerable disagreement by Turnbull and Hoffman [56] who studied self-diffusion of silver bicrystals by x-ray analysis and arrived at opposite conclusions about the diffusional effectiveness of isolated tubes of separate dislocations. As in the experiment of Smoluchowski, the silver bicrystals had a common axis of rotation coinciding with the [100] direction and measurements were made in the range of angles from 9 to 28°. They found that the self-diffusion coefficient along the dislocational tubes, D_p, is independent of θ for $\theta \leq 28°$ and at 500°C is larger by several orders than the coefficient of volume diffusion. They noted a considerable dependence of the depth of penetration on the orientation of the boundary in the case of $\theta = 9°$, which result surprised the authors.

Thus according to experimental data [56] isolated dislocational tubes are effective for diffusion, contrary to the assertions of Smoluchowski.

Apparently Turnbull and Hoffman [56] were correct in stating that Smoluchowski's method was insufficiently precise to reveal preferential diffusion along the grain boundaries. In connection with this it should be noted that Smoluchowski investigated diffusion at somewhat higher temperatures than Turnbull and Hoffman and it is now clearly demonstrated that preferential diffusion along the boundaries predominates over volume diffusion only at low temperatures while at high temperatures the transfer of atoms along the boundaries occurs at the same rate or a slightly different rate from the transfer through the lattice. Thus Leymonie and Lacombe [62] studied preferential self-diffusion along the grain boundaries of iron and found that at 700°C the depth of penetration along the boundary was 10 times higher than along the grains; at 800°C the ratio of the depth of penetration in the two cases was only 2 while at 850°C the depth of penetration were almost equal. Consequently the decrease of the angle as well as the increase of the temperature will decrease the precision of the measurement of preferential diffusion along the boundaries. In the light of these results let us consider Smoluchowski's result [55] concerning the diffusion of zinc into copper, where he obtained the following values for the critical angle θ_s:

At a diffusion temperature of 550° ... $\theta_s = 9°$
593° ... $\theta_s = 18°$
649° ... $\theta_s = 23°$.

Considering the rapid decrease of the angle θ_s with decreasing temperature, it seems reasonable to expect that at temperatures close to 500°C the value of the critical angle will be zero, i.e., preferential diffusion along the boundaries will occur at any small angle. Consequently, Smoluchowski's assertion that there is no diffusion along the dislocational tubes separated from each other by dislocations is apparently erroneous.

Lately Hoffman [63] has published an investigation of the anisotropy of diffusion along dislocational tubes. Hoffman started with the assumption that the coefficient of self-diffusion along the edge dislocations, $D_{\parallel}$ is several orders higher than for self-diffusion along the lattice, while the coefficient of self-diffusion perpendicular to the dislocational tubes $D_{\perp}$ must depend on the distance between dislocations, i.e., on the disorientation angle of the grains. Thus if the Burgers-Bragg dislocation model is correct [12] there must exist an anisotropy of diffusion depending on the disorientation angle θ of the grains. Hoffman [63] measured the self-diffusion of silver along the [100] axis of disorientation of two grains of a bicrystal and along the direction perpendicular to this axis as a function of the disorientation angle. Using the analysis published by Fisher [50], he calculated the coefficient of diffusion for different disorientation angles. Figure 18 represents the ratio $D_{\parallel}/D_{\perp}$, as a function of the disorientation angle. Because the value of $D_{\perp}$ is so small it was impossible to calculate the ratio for $\theta < 16°$. However, this fact in itself constitutes a qualitative proof of high anisotropy of diffusion at small disorientation angles. Figure 18 shows that the anisotropy of diffusion is preserved in boundaries with maximum disorientation, $\theta = \sim 45°$. Thus the as-

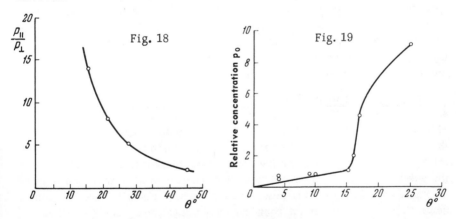

Fig. 18. Variation of the anisotropy of self-diffusion of silver as a function of the disorientation angle of grains in a bicrystal.

Fig. 19. Variation of the equilibrium concentrations of polonium atoms in boundaries of bicrystals of a Pb—Bi alloy as a function of the disorientation angle, θ.

sumption that boundaries with large angles can be represented as homogeneous masses of distorted material was not confirmed in this investigation.

Thus the rate of preferential diffusion along the grain boundaries depends on the disorientation angle of grains within a relatively wide range of angles, including boundaries with small and large angles: there is no doubt about this experimental result since it is confirmed by the general concept that diffusion depends on the structure and therefore the change in the structure of the boundary due to the change in the angle must lead to variation in the diffusion rate. The investigation by Smoluchowski on one hand and Turnbull and Hoffman on the other gives rise to the expectation that the study of preferential diffusion along the boundary will result in very valuable data for the construction of the model of the structure of the boundary. The phenomenon of the anisotropy of diffusion along the boundaries and its dependence on the angle θ make it possible not only to verify the dislocational model of the boundary but also to consider the structure and properties of "large-angle boundaries" from the dislocational viewpoint, since it has become clear that considerable anisotropy of diffusion is also found in boundaries with maximum disorientation angles.

OTHER PROPERTIES OF BOUNDARIES DEPENDING ON THE DISORIENTATION ANGLE OF THE GRAINS

During recent years the results of measurements of some properties of boundaries as functions of the disorientation angles of grains have been published. As far as we know there have been few such investigations and they give insufficient data for definite conclusions about the details of the structure of the boundary. Nevertheless it is interesting to consider some of them briefly, at least to see to what degree they confirm the general assumption of the dependence of the structure on the disorientation angle and whether they provide new data on the details of the structure.

Segregations of Impurities on the Boundaries. Thomas and Chalmers [64] studied the segregations of polonium along the boundaries of a lead—bismuth alloy by the radiographic method. They found a relationship between the equilibrium concentration of polonium atoms and the disorientation angle θ of neighboring grains of the bicrystal, which is shown in the curve of Fig. 19. The curve shows that boundaries with small and large angles have different effects on segregation. Thus for small angles the concentration of polonium in the boundary increases (almost linearly) with increasing angles. Above $A = 15°$ the rate of increase in the concentration of polonium increases sharply with the angle. This experimental fact confirms the assumption that the structure of boundaries with small and large angles of disorientation is different, which is in accord with the dislocational theory of the structure of grain boundaries (according to the theory, in boundaries with small angles dislocations are isolated while in boundaries with large angles the dislocations assemble in groups and form areas with greatly distorted structures).

Melting of Boundaries. The phenomenon of preferential separation along the brain boundary at temperatures close to the melting point is usually called the melting of boundaries. Apparently Chalmers [8] was the first to study this phenomenon as a function of the disorientation angle of grains in tin bicrystals of different degrees of purity (99.89, 99.996, 99.986%). The experiments consisted of slowly increasing the temperature of the bicrystalline sample of the required shape and orientation of the grain boundaries. At the same time the sample was subjected to tensile stress perpendicular to the grain boundary. The temperature at which the two grains of the bicrystal separated from each other was determined. On the basis of these measurements Chalmers asserted that the temperature at which the two grains separate is 0.14°C lower than the melting point of the basic material. This decrease of the melting point is independent of the value of the applied stress, the rate of heating, and the disorientation angles of the grains within the range of 14-85°, i.e., boundaries with large angles. Twinning boundaries did not show any lowering of the melting point.

Recently Weinberg and Teghtsoonian [65] made a detailed study of the melting of grain boundaries as a function of applied stress, rate of heating, concentration of impurities, and the disorientation angle of boundaries from 5 to 80°. They found that in samples with boundaries having small angles of disorientation ($\theta < 12°$) the grains did not separate along the boundary but rupture usually occurred in cross sections unrelated to the boundary of the bicrystal. In bicrystals having a boundary with a large angle of disorientation rupture always occurred along the boundary. The temperature at which rupture occurred was independent of the applied stress and rate of heating and was equal (within the limits of precision of the experiment, 0.02°C) to the melting point of the basic material. The separation of grains of a bicrystal occurred not instantaneously the moment the melting point was reached but occurred over a certain length of time (see the plateau on the heating curve) which depended on the experimental conditions, which in turn depended on the disorientation angle, the applied stress, and the degree of purity of the material. An increase in the amount of impurities in the material affected the melting point of both the boundary and the base metal according to the state diagram. Analogous results were obtained for aluminum bicrystals.

The most interesting finding in this investigation is the difference in the behavior of boundaries with small and large angles. Furthermore, the value of the critical angle at which the behavior of the boundaries changes ($\theta = 12°$) is in accord with the data on the energy of boundaries in tin given by Chalmers and Aust [38]. From this data Read [31] calculated that the value of θ_m

equals 12.2°, i.e., the angle of disorientation corresponding to the maximum energy of the boundary. The identity of these two angles indicates that the structure of the boundaries or, more precisely, their deformation energy, plays an important role in the melting of the boundary. However, at the present time there are no reasons to negate a certain effect of the impurity atoms segregated on the boundary. The fact that in the case of boundaries with small angles the samples do not rupture preferentially along the boundary can be explained (from this point of view) as a low degree of segregation of impurities on the boundary, as was shown by Thomas and Chalmers [64] (see Fig. 18).

Macromotion Along the Grain Boundaries. Some years ago Chalmers et al.[66], conducted experiments on the relative slip of two grains of a tin bicrystal along the boundary. It turned out that at a temperature of 220°C and a tangential stress τ of 590 g/cm^2, the total slip during 50 hr was 0.1 mm. This result has been confirmed in other investigations [67, 68].

Recently Tung and Maddin [69] published results concerning the relationship between the relative slip along the boundary and the disorientation angle in aluminum bicrystals (99.99% Al). The bicrystalline samples consisted of grains with a common axis [110] and consequently the boundaries were purely rotational boundaries with angles in the range of 20-85°. The creep stress along the boundary was chosen in the range of 30-75.5 g/cm^2 and the study was made at temperatures of 450-550°C. The creep curve under all stresses and at all temperatures had a cyclic character (the creep rate varied periodically). This type of curve was found earlier by Chang and Craut [70] in the case of large-grained aluminum, where the intermittent character of the creep process was actual and not averaged from a large number of separate processes on different grains.

Generally the creep curve obtained by Tung and Maddin had an inhibition period, then a linear region, and then alternating regions of increased and decreased rates.

Figure 20 represents typical slip along a horizontal boundary. The value of relative slip was measured by the distance between scratches made on the bicrystal perpendicular to the boundary. The variation of the activation heat as a function of the disorientation angle was calculated from the straight line representing the variation of the logarithm of the rate at the beginning of creep as a function of $1/T$, where T is absolute temperature. Figure 21 represents the variation of the activation heat (the dotted part of the curve is extrapolated) as a function of the disorientation angle θ. The curve shows that the activation heat increases with increasing angles θ reaching a maximum when $\theta = 90°$.

Fig. 20. Macroslip along the grain boundaries of aluminum bicrystals (×200). The vertical bands are traces of scratches.

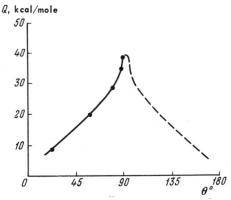

Fig. 21. Variation of the activation heat of a macroslip along the boundaries as a function of the disorientation angle of grains, θ.

These authors assumed that n in the Mott formula [see formula (14)] must depend on the disorientation angle since it represents a value characterizing the structure of the boundary. Assuming in the case of aluminum that $L = 2.55$ kcal/mole and taking the activation heat Q from the curve in Fig. 21 (and neglecting T/T_m with respect to unity, as Mott did), they obtained the following values of n as a function of the disorientation angle:

θ		*n*
20°	. . . .	. 3.4
55°	. . . .	. 7.9
80°	. . . .	11.5
85°	. . . .	13.5
88°	. . . .	15.5

Thus n = 14 only for boundaries with disorientation angles between 85 and 88° while Mott assumed that n = 14 for boundaries with angles greater than 10-15°.

The intermittent character of the process of creep during macroslip leads us to assume that the process of macromotion along the grain boundary occurs as the result of two processes: slip along the islands of undistorted structure, and self-diffusion, which renders the areas of large discontinuities more orderly.

It should be noted that the activation heat increases with the disorientation angle because the structure of these distorted areas become more complex. It is interesting that Aust, Harrison, and Maddin [71], in an investigation of the migration of boundaries of bicrystals as a function of the disorientation angle θ, found that the activation heat of migration was Q = 43 kcal/mole at θ = 55° and Q = 20 kcal/mole at θ = 85°, i.e., the activation heat of migration decreases and the activation heat of macroslip along the boundaries increases with increasing disorientation angles.

In summary let us note that most properties of the grain boundaries depend on the disorientation angle of the boundary, i.e., on the structure of the boundary. Therefore in order to determine the role of boundaries in different phenomena (creep, chemical properties, brittleness, etc.) it is necessary to know the form of the dependence of the properties on the angles and take into account the sharp difference in the behavior of boundaries with small and large disorientation angles of grains.

CONCLUSION

Let us briefly summarize the most salient results of this review. The discovery and investigation of polygonization and the study of the structure and properties of subgrains provided a large amount of experimental data whose generalization has extended our knowledge of the structure and properties of grain boundaries. Discrete etching, which revealed the discrete structure of the subboundaries, has been of decisive value in these investigations. The most important result of discrete etching was the finding that the shape of the etching figures on subboundaries and subgrains is the same, i.e., the same selective etching agent reveals the etching figures on the crystal surface of subgrains and subboundaries. There are convincing proofs that these etching figures do not reveal surface defects but rather the defects which extend from the body of the crystal to the surface. It is interesting that there is a strict orientational relationship between the slip planes and the polygonization subboundaries; this indicates that the defects from which the subboundaries are built are defects which migrate from the slip planes.

Aside from these experimental investigations of the substructure, theoretical concepts of the structure of the boundary were developed from the dislocation viewpoint; this theoretical work led to a number of quantitative relationships [Eqs. (1), (8), and (9)] for boundaries with small angles and new phenomena were predicted on the basis of it: motion of boundaries with small angles, orientational dependence of subboundaries and slip planes, dependence of the anisotropy of diffusion on the disorientation angle, etc.

These concepts led to a number of experiments concerned with quantitative checking of Eqs. (1), (8), and (9) and to the search for the predicted phenomena. Experimental confirmation of Eq. (1) revealed the physical meaning of the dependence of the structure of boundaries on the angle, i.e., it showed that the density of dislocations in the boundary depends on the angle of the boundary.

Measurements of the energy of the boundary as a function of the disorientation angle of the grains and checks of relationships (8) and (9) by independent experiments gave quantitative confirmation of the exactness of the dislocational model of boundaries with small angles.

The discovery of the predicted motion of grain boundaries under the effect of applied stress not only reconfirmed the dislocational model of boundaries with small angles by an independent method but also revealed a number of particularities of the structure of grain boundaries, such as the combination and separation of boundaries of the same and different signs at any (high or low) temperature, which indicates the discrete structure of boundaries with small angles.

The dependence of the rate of preferential diffusion along grain boundaries and also the anisotropy of diffusion as a function of the disorientation angle of grains not only confirm the dislocational model of boundaries with small angles but, more important, give data making it possible to describe the structure of boundaries with large angles more precisely. The dislocational model of grain boundaries predicts (and experiments confirm) a difference in the behavior of boundaries with small angles (constructed from noninteracting dislocations) and boundaries with large angles (interacting dislocations). Therefore the dependence of the properties of boundaries on the degree of freedom can change drastically and is different for regions with small and large angles. Therefore it is impossible to clarify the role of boundaries in different phenomena in polycrystals, where the role of boundaries with small and large angles cannot be differentiated. Such experiments are possible only in the case of bicrystals, tricrystals, and samples with columnar structure or texture, for which the boundaries can be characterized to some extent by their parameters, i.e., by their disorientation angles.

Thus the dislocational model of grain boundaries has been fruitful, making it possible to predict a great number of phenomena and derive quantitative relationships. This model has made it possible to investigate qualitatively, and in some cases quantitatively, the most important properties of boundaries as a function of the disorientation angle of the grains. It can be assumed that in the near future we shall derive quantitative as well as qualitative relationships for all the properties of boundaries as a function of the disorientation angle.

LITERATURE CITED

1. S. J. Konobeevski, and I. Mirer, "Die röntgenographische Bestimmung elastischer Spannungen in gebogenen Kristallen," Z. Kristallogr. 81, 69 (1932).
2. C. Crussard, "Etude du recuit de l'aluminium," Rev. métallurgie 41, 133 (1944).
3. P. Lacombe and L. Beaujard, "The application of etch figures on pure Al (99.99%) to the study of some micrographic problems," J. Inst. Metals 74, 1 (1947).
4. R. W. Cahn, "Recrystallization of single crystals after plastic bending," J. Inst. Metals 76, 121 (1949).
5. C. G. Darwin, "The theory of x-ray reflection," Phil. Mag. 27, 315 (1914).
6. F. Hargreaves and R. J. Hills, "Work-softening and a theory of intercrystalline cohesion," J. Inst. Metals 41, 269 (1929).
7. B. Chalmers, "The influence of the difference of orientation of two crystals on the mechanical effect of their boundary," Proc. Roy. Soc. A 162, 120 (1937).
8. B. Chalmers, "Crystal boundaries in tin," Proc. Roy. Soc. A., 175, 100 (1940).
9. B. Chalmers, "Some crystal-boundary phenomena in metals," Proc. Roy. Soc. A. 196, 64 (1949).
10. M. V. Klassen-Neklyudova and T. A. Kontorova, "The nature of intercrystalline layers (review)," Usp. Fiz. Nauk 22, No. 3, 249 (1939).
11. R. King and B. Chalmers, "Grain boundaries," Progr. Metal phys. 1, 126 (1949).
12. J. M. Burgers, "Geometrical considerations concerning the structural irregularities to be assumed in a crystal," Proc. Phys. Soc. 52, 23 (1940).
13. W. T. Read and W. Shockley, "Dislocation models of crystal grain boundaries," Phys. Rev. 78, 275 (1950).
14. J. S. Ke, "Experimental evidence of the viscous behavior of grain boundaries in metals," Phys. Rev. 71, 533 (1947).

15. N. F. Mott, "Glide at grain boundaries and grain growth in metals," Proc. Phys. Soc. 60, 391 (1948).
16. P. Lacombe and N. Yannaquis, "Subboundary and boundary structures in high-purity aluminum," Report of the Conference on Strength of Solids, Bristol, 1947 (Phys. Soc. London, 1948).
17. F. A. Rohrmann, "The effect of heat treatment on the corrosion of high-purity aluminum," Trans. Electrochem. Soc. 66, 229 (1934).
18. H. Mahl and L. N. Stranski, "Über Atzfiguren an Al-Kristalloberflächen," Zschr. phys. Chem. 52, 257 (1942).
19. C. M. Tucker, "New etching reagent for the macrography of aluminum and its alloys," Metals and Alloys 1, 655 (1930).
20. C. S. Barrett and L. H. Levenson, "The structure of aluminum after compression," Trans. AIME 137, 112 (1940).
21. W. Shockley and W. J. Read, "Quantitative predictions from dislocation models of crystal grain boundaries," Phys. Rev. 75, 692 (1949).
22. F. L. Vogel, W. G. Pfann, H. E. Corey, and E. E. Thomas, "Observations of dislocations in lineage boundaries in germanium," Phys. Rev. 90, No. 3, 489 (1953).
23. F. L. Vogel, "Dislocations in low-angle boundaries in germanium," Acta metallurg. 3, No. 3, 245 (1955).
24. S. Amelinckx, "Etch-pits and dislocations along grain boundaries, slip lines and polygonization walls," Acta metallurg. 2, No. 6, 848 (1954).
25. W. G. Pfann and L. C. Lovell, "Dislocation densities in intersecting lineage boundaries in germanium," Acta metallurg. 3, No. 5, 512 (1955).
26. W. G. Pfann and F. L. Vogel, "Observations on the dislocation structure of germanium crystals," Acta metallurg. 5, No. 10, 377 (1957).
27. J. J. Gilman and W. G. Johnston, "Observations of dislocation glide and climb in lithium fluoride crystals," J. Appl. Phys. 27, No. 9, 1018(1956).
28. G. Wyon and C. Crussard, "Modifications de structure de l'aluminium au cours du fluage," Rev. metallurgie 48, 121 (1951).
29. G. Wyon and P. Lacombe, "The influence of dislocations and impurities on the distribution and size of etch figures on pure aluminum," Report of the Conference on Defects in Crystalline Solids, Bristol, 1954 (Phys. Soc. London, 1955).
30. J. C. Suits and R. G. Low, "Dislocation etch-pits in silicon iron," Acta metallurg. 5, No. 5, 285 (1957).
31. W. J. Read, Dislocations in Crystals (1953) ch. 14.
32. J. Washburn and E. R. Parker, "Experiments on the stress-induced motion of small-angle boundaries," Trans. AIME 194, 1976 (1952).
33. C. H. Li, E. H. Edwards, J. Washburn, and E. R. Parker, "Stress-induced movement of crystal boundaries," Acta metallurg. 1, 233 (1953).

34. S. N. Buckley and K. M. Entwistle, "The Bauschinger effect in super-pure aluminum single crystals and polycrystals," Acta metallurg. 4, No. 4, 352 (1956).

35. D. W. Bainbridge, Chon-Hsien Li, and E. H. Edwards, "Recent observations on the motion of small angle dislocation boundaries," Acta metallurg. 2, No. 2, 322 (1954).

36. N. F. Mott, "Mechanical properties of metals," Proc. Phys. Soc. B, 64, 729 (1951).

37. C. S. Smith, "Grains, phases and interfaces: an interpretation of microstructure," Trans. AIME 175, 15 (1948).

38. B. Chalmers and K. T. Aust, "The specific energy of crystal boundaries in tin," Proc. Roy. Soc. A, 201, 210 (1950).

39. B. Chalmers and K. T. Aust, "Surface energy and structure of crystal boundaries in metals," Proc. Roy. Soc. A. 204, 359 (1950).

40. C. G. Dunn, "Controlled grain growth applied to the problem of grain boundary energy measurements," Trans. AIME 185, 72 (1949).

41. C. G. Dunn, "The effect of orientation difference on grain boundary energies," Trans. AIME 185, 125 (1949).

42. C. G. Dunn, F. D. Daniels, and M. J. Botton, "Relative energies of grain boundaries in silicon iron," Trans. AIME 188, 1245 (1950).

43. A. P. Greenough and R. King, "Grain boundary energies in silver, J. Inst. Metals 79, 415 (1951).

44. J. Sawai and M. Nishida, Z. anorg. und. allgem. Chem. 190, 375 (1930).

45. R. Shuttleworth, Discussion of the article by J. C. Fisher and C. G. Dunn, "Surface and interfacial tensions of single-phase solids," Imperfections in Nearly Perfect Crystals (1952) p. 343.

46. L. H. Van Vlack, "Intergranular energy of iron and some iron alloys," Trans. AIME 191, 251 (1951).

47. H. Brooks, "Theory of internal boundaries," Metal interfaces ASM (1952) p. 20.

48. J. C. Fisher and C. G. Dunn, "Surface and interfacial tensions of single-phase solids," Imperfections in Nearly Perfect Crystals (1952) p. 317.

49. R. M. Berrer, Diffusion in Solids [Russian translation] (Il, 1948).

50. J. C. Fisher, "Calculation of diffusion penetration curves for surface and grain boundary diffusion," J. Appl. Phys. 22, No. 1, 74 (1951).

51. R. J. P. Whipple, "Concentration contours in grain boundary diffusion," Phil. Mag. 45, 1225 (1954).

52. P. A. Beck, P. R. Sperry, and H. Hu, "The orientation dependence of the rate of grain boundary migration," J. Appl. Phys. 21, 420 (1950).

53. M. R. Achter and R. Smoluchowski, "Diffusion in grain boundaries and their structure," J. Appl. Phys. 22, 1260 (1951).

54. M. H. Dawson, "Diffusion in grain boundaries," J. Appl. Phys. 23, 373 (1952).

55. R. Flanagan and R. Smoluchowski, "Grain boundary diffusion of zinc in copper," J. Appl. Phys. 23, 785 (1952).

56. D. Turnbull and R. E. Hoffman, "The effect of relative crystal boundary orientations on grain boundary diffusion rates," Acta metallurg. 2, 2, No. 3, 419 (1954).

57. S. R. L. Couling and R. Smoluchowski, "Anisotropy of diffusion in grain boundaries," J. Appl. Phys. 25, 1538 (1954).

58. M. R. Achter and R. Smoluchowski, "Anisotropy of diffusion in grain boundaries," Phys. Rev. 83, 160 (1951).

59. C. W. Haynes and R. Smoluchowski, "Grain boundary diffusion in a body-centered cubic lattice," Acta metallurg. 3, No. 2, 130 (1955).

60. C. G. Dunn and G. C. Nonken, "Production of oriented single-crystal silicon iron sheet," Metal Progr. 64, No. 6, 71 (1953).

61. L. Vassamillet, Cited in: R. Smoluchowski, Report of the Conference on Defects in Crystalline Solids, Bristol, 1954 (Phys. Soc. London, 1955).

62. C. Leymonie and P. Lacombe, "Autodiffusion preferentielle dans les joints de grains de fer cubique centre," Acta metallurg. 5, 115 (1957).

63. R. E. Hoffman, "Anisotropy of diffusion in grain boundaries," Acta metallurg. 4, No. 1, 97 (1956).

64. R. W. Thomas and B. Chalmers, "The segregation of impurities in grain boundaries," Acta metallurg. 3, 17 (1955).

65. F. Weinberg and E. Teghtsoonian, "Grain boundary melting," Acta metallurg. 5, 455 (1957).

66. R. King, R. W. Cahn, and B. Chalmers, "Mechanical behavior of crystal boundaries in metals," Nature 161, 682 (1948).

67. G. Chaudron, P. Lacombe, and N. Yannaquis, "Sur le comportement des joints de grains au cours du processes de fusion de l'aluminium tres pur," Compt. rend. 226, 1372 (1948).

68. R. King and R. A. E. Puttick, Report Met. 45.

69. S. K. Tung and R. Maddin, "Shear along grain boundary in Al-bicrystals," J. Metals 9, No. 7. Sec. 2, 905 (1957).

70. H. C. Chang and N. J. Craut, "Observations of creep of the grain boundary in high-purity aluminum," Trans., AIME 194, 619 (1952).

71. K. T. Aust, E. H. Harrison, and R. Maddin, "Observations on grain boundary migration in aluminum bicrystals," J. Inst. Metals 85, No. 9, 15 (1956).

72. S. Amelinckx, "The geometry of grain boundary junctions," Physica 23, 663 (1957).

73. S. Amelinckx and W. Dekeyser, "The structure and properties of grain boundaries," Solid State Physics 8 (1959).

74. J. Friedel, B.D. Cullety, and C. Crussard, "Study of the surface tension of grain boundary in metals as a function of the orientation of the two grains which the boundary separates," Acta Metallurgica 1, No. 1, 79 (1953).